The Art of INFORMATION WARFARE

Insight into the Knowledge Warrior Philosophy

Richard Forno

Ronald Baklarz, CISSP

Universal Publishers/UPUBLISH.COM
1999

To learn more about this book, its authors, speaking engagements, or book signings, visit the official home page of *The Art of Information Warfare* at www.taoiw.org.

Cover Art and Layout by IMAGIC Design, Dunkirk, MD. and can be found at www.imagicduo.com.

ISBN: 1-58112-857-6

Universal Publishers / UPUBLISH.COM

Praise for
THE ART OF INFORMATION WARFARE

"Provides a lucid evaluation clarifying beyond question that we are now engaged in technological warfare…the book is serious while entertaining, understandable while clear and concise…Forno and Baklarz provide solutions."

James Kaljian
Chairman, Criminal Justice Department
Valley Forge Military College

"A must read for any one who knows or who thinks they know all there is to know about Information Warfare. Though the current "buzz-word" is Information Operations (IO), this book discusses these and other related topics in a way that most anyone can understand and grasp their concepts and principles…. I highly recommend *The Art of Information Warfare* definitely be made part of any security professionals' permanent library."

Stephen Glennan, OCP
President
Operations Security Professionals Society

"Excellent source material that Corporate Security Managers should make part of their Information Security Education program."

Mark McGovern
President, Mid-Atlantic Chapter
High Technology Crime Investigations Association

"Very neat using the sword, mirror, and jewel as metaphors for Information Warfare….is it a two-edged sword? We seem to be selling our best capabilities to anyone who wants it, and will regret it later when we must defend ourselves against our own technology and might get cut by our own sword!"

US Air Force Colonel
Eglin AFB FL

"Provides a unique and unconventional view of information warfare…using the 2,500 year old principles and strategies of *The Art of War* as their construct, the authors have taken the precepts of Sun Tzu and applied them to today's new battlefield of the Information Warrior."

Donald Withers
Vice President
WarRoom Research, Inc.

Information drives military, national, corporate, and individual decision-making processes. Thus, Information Warfare seeks to deny, disrupt, or modify this process for the purposes of enabling an adversary to achieve their goals through the use of attacks (cyber, kinetic, psychological or psychotronic actions) directed against his opponent's decision-making abilities or infrastructures; or by actions (such as deception or disinformation) directed against the actual information with the ultimate goal of adversely affecting the outcome of the opponent's decisions and subsequent actions.

Author's Working Definition

Dedication

The authors of The Art of Information Warfare *wish to dedicate this book to the elite cadre of men and women—military and civilian—who are the front-line cyberspatial defenders and protectors of the networked society. Their devotion to duty, willingness to respond to late-night pager calls, patience to work with and tolerate upper management, and perseverance to parse, sort, and analyze reams of system and network traffic logs during an investigation make them the true guardians of the Information Age.*

We've been there, done that, and loved every minute of it.

Forward

"What is called foreknowledge cannot be elicited from spirits, nor from gods, nor by analogy with past events, nor from calculations. It must be obtained from men who know the enemy situation." —Sun Tzu

CONTRARY TO POPULAR BELIEF, Information Warfare is not a new concept.

In fact, the very nature of Information Warfare is as classic as the art of war itself, but it is all too common to view information warfare as a new type of war (the current "revolution" versus "evolution" in military affairs debate). However, the underlying concept of information warfare remains unchanged—that being to *deny the enemy the ability to wage war by depriving him of his will and capability to fight against you.* Whether the final combat occurs through a conventional attack, information-based or precision-guided weapon, the end goal is the same. To deprive the enemy of his will and capability to fight.

Unfortunately, the rash of computer-related security incidents and the glamour of technology has somewhat sensationalized the study of this very real concern to our telecommunications, economic, social, resources, and government circles. However, the threat of information— or infrastructure—warfare's precision attacks on particular strategic resources affecting a large number of people or social services such as power grids or water supplies ranks quite high on the government's Most Serious Threat List along with terrorism, chemical or biological warfare, drug trafficking, and organized crime.

As you read the following pages, think of how you may enhance your information security posture through both traditional and nontraditional methods. Also reflect on your inherent vulnerabilities as a member of a "wired" organization and society as both information professional and private citizen. Remember that many of the philosophies, observations, and tactics in this book are unconventional and "out of the box" approaches to the issue—just like an adversary would develop in an attack against you!

And remember that there are other facets of Information Warfare that do not require a computer or modem to be used in the attack for them to be effective.

Information Warfare is only now becoming a national concern by many government and business organizations around the world from both the law enforcement and national security perspectives. While we have not arrived at the "too little too late" stage, it is imperative to develop a cohesive, proactive strategy to deal with this established threat and the attendant issues associated with a decentralized adversary, one that has no geographic confines and affords the target no indication or warning system as was established during the Cold War. Dealing with these new threats requires *a new way of thinking, a new way of doing things* in business and government organizations, and a *conceptual realization that there exists increasingly complex relationships* of the many mission-critical systems on the planet. The speed that things happen in the Information Age also adds to the complexity—*isolationism, turf wars, or time-consuming bureaucratic wrangling is not the answer*. Indeed, No Man or Organization Is An Island...we are all connected in some way to somebody somehow.

Rick wishes to thank COL John Macartney, USAF (Retired) and Zhi Hamby for giving him his start in the "profession"; the Operations Security Professionals Society; faculty advisors Agnes Campbell, Bob Byrne, Wade Stallard, and Pat Murray; the many information warriors he's worked and interacted with in the course of his career and this book (particularly Rod and Danny); his close friends and family, and especially Theresa, who has been constantly in his thoughts during the literary process.

Ron wishes to thank his wife Suzann and family—especially Mom and Dad for "hard wiring his kernel," the Information Science Department at the University of Pittsburgh—Dr. Toni Carbo and Dr. James Williams, all the folks in the Naval Nuclear Program who provided his firm roots in INFOSEC—Harvey Rosenblum, Gene Olson, Tim Glock, Mike Fillipiak, and John Todd, the great people at the U.S. House of Representatives especially John W. Lainhart, IV, Danny Gottovi, Rod Murphy, and all those yet-to-be-developed relationships in the vast security arena.

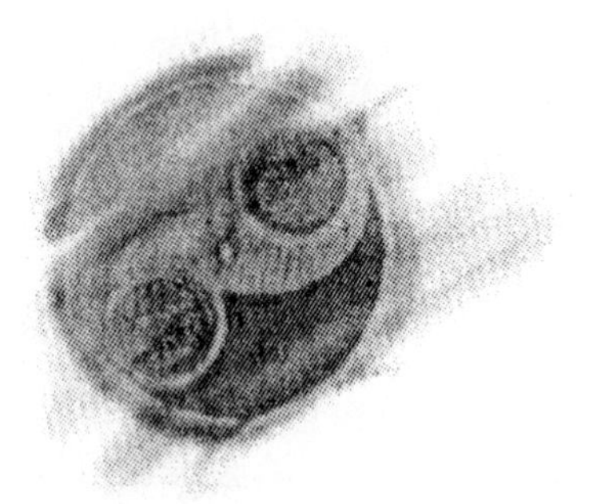

Why the Asian Metaphor?

THE MEANING OF THE ABOVE SYMBOL—while simple in appearance—is quite philosophical. Its origin comes from the oriental philosophy called *Eum-Yang*; in Chinese, it is pronounced "Yin-Yang" and hence the English version of the word. In Korea, the symbol of Yin and Yang is called *Tae Guk* and summarizes the thoughts of the I *Ching*, or "Book of Changes." *Yin* means "dark and cold" while *Yang* means "bright and hot."

A very old Chinese book called the *Choo-Yuk* claims that all objects and events in the world are expressed by the movement of Yin and Yang. For example, the moon is Yin while the sun is Yang. The earth is Yin and the sky is Yang. The night is Yin and the day is Yang. The winter is Yin and the summer is Yang. Discipline is Yin, and recklessness is Yang. Good may be the Yin and evil the Yang. Thus, Yin and Yang are relative terms.

It is through this Asian metaphor that we chose to serve as the setting for our discussion of Information Warfare. Information is a two-edged sword, with the Yin being the ability to use such information to help and strengthen people and organizations and the Yang being the dark actions of wrongfully using information to hurt people and damage organizations. Just as the *I Ching* guides us through an ever-changing but balanced life, so will *The Art of Information Warfare* provide guidance in the dynamic environment of cyberspace to those in seeking balanced approaches to cyberspatial security.

(more)

Part One—*White Belt Theory*—imparts to the reader the theory and insight required to fully understand the fundamentals (the Yin and Yang) of Information Warfare as a person, member of an organization, and ultimately as a society. White signifies a birth, or beginning, of a seed, and thus the white belt student is a beginner searching for knowledge of the *Art*. Part Two—*Black Belt Knowledge*—is divided into numbered sections (called *Poomse*, pronounced "Poom-Say" and meaning "style forms") that focus on a particular task required by the professional Information Warrior to conduct his activities in the Information Warfare arena. Black signifies the darkness beyond the Sun and thus the black belt seeks new knowledge and new innovations in the *Art*. As he/she begins to teach others and innovate, he/she begins planting new seeds—their students and those they mentor—many of whom will take root deep into the *Art*, blossom and grow through the ranks in a never-ending process of self-growth, knowledge, and enlightenment.

It is through this and no other way that the Novice becomes a Knowledge Warrior.

Table of Contents

INTRODUCTION

It is not a matter of if but when. . .

Recent media releases have been prolific in their rhetoric regarding information security. The explosion of the Internet, the World Wide Web, and attendant technologies has fueled a flurry of concern on the part of our nation's top security experts, company CEOs, Military brass, and government officials.

In 1998, the Computer Security Institute and the Federal Bureau of Investigation conducted a survey of information security managers. Of the over five hundred respondents:

64% reported computer security breaches.
54% were attacked via Internet connections.
70% reported unauthorized uses by insiders.

In May 1996, the General Accounting Office reported that the Pentagon computer systems were attacked some 250,000 times with a 65% success rate. The GAO report stated that while many of these attacks were unobtrusive and involved unclassified networks only, many resulted in theft or destruction of sensitive-but-unclassified data costing "tens of millions of dollars."

You may not even be aware of the fact that we are engaged in this silent war. This is because cyberwar or information warfare is not fought with guns, bombs and armies; it is fought by warriors armed with knowledge, computers and programs connected to the incomprehensible web we call the Internet. Today's war is fought for, around, and in the silicon trenches, fiber optic highways, and in the ether that has and is continuing to cloud our quickly changing cybersociety. Today's enemy does not march across the open countryside, nor hunker down in a dark foxhole. Today's enemy mounts his attack from the comfort of his own domain or even from his own home, taking the form of a disgruntled employee, a fourteen-year-old "hacker," a cyber-thief, or agent of a hostile government or military. Silent skirmishes can now take place at the bit-level; in the deepest recesses of an operating system where the inadvertent setting of a "sticky bit" can leave one's fortress open to an unauthorized access.

Information warfare even includes its own version of biological warfare. This aspect of warfare takes on the form of computer viruses and other types of malicious code. Electronic virii replicate and cause damage with the mortality rate equal to or worse than the equivalent biological Ebola. Ambushes are encountered along the thousands of miles of wire that can be exploited by "session highjackers." In a more classic, Eastern description, network routers can be "spoofed" into letting an intruder disguised as one of the "villagers" into the "castle."

A daily battle is being waged, regardless of whether one has a laptop computer, desktop computer, workstation or mainframe. There is little difference if one is attached to a local area network, wide area network, or a mom-and-pop Internet service provider. As with many other facets of modern life, hostile forces lurk around every corner. Recent publicized attacks range from the manifestation of viral code to a hack against an organization's Internet firewall or World Wide Web site. The bottom line is that these opposing forces will cost large sums to defend against and/or react to.

The summer of 1996 was witness to hackers who broke into the World Wide Web sites of the Department of Justice and the Central Intelligence Agency. The perpetrators replaced the content of the web sites with their own pornographic "information." The embarrassment caused by these incidents will not soon be forgotten. Your investment in hardware, software, data, and information is expensive, as are the costs to generate, load, and then maintain it. As we know, an intrusion into a system can result in a denial of service, loss of data, or. . . worse.

However, times are changing, and the record is starting to show a shift in computer security-related incidents. Several organizations are conducting ongoing quantitative research into the true scope of the level of vulnerability in American corporations. After 18 months of interviews with 320 Fortune 1000 firms, the 1998 WarRoom Research report

"Corporate America's Competitive Edge: An 18 Month Study into Cyber-security and Business Intelligence Issues" continues where its 1996 U.S. Senate report left off. . . with some compelling new information. The report finds that **63% of those interviewed now indicate that the most damaging threat to their information resources comes from *outside* the organization—*not from insiders*** as has been reported during the past few years, including the famed annual FBI/CSI reports.

That being said, what needs to be understood are the tremendous motivations for conducting Information Warfare activities. These motivations include but are not limited to:

(1) **Information Warfare is cheap** when compared to other types of attacks. According to Moore's Law, computing speed and memory capacities double approximately every 18 months and therefore the cost of computing capability such as hardware and most software continues to fall dramatically. One must also remember that it does not take a tremendous amount of computing power to conduct Information Warfare activities. Low-end technology, supplies, software and the traditional techniques of conventional warfare such as deception, psychological operations, or kinetic attacks (bombing) are all that is required.

(2) **Information Warfare is less risky** to an adversary compared to other types of attacks. This will become more significant as America takes action to circumvent the threats associated with physical forms of attacking the information infrastructure. Cyber terrorists will soon discover that logical warfare can be just as effective and debilitating as traditional terrorism or kinetic warfare—however, the chances of getting caught or injured are dramatically reduced given the "virtual" nature of such attacks.

(3) **The odds of getting caught are very low.** Discovering attacks or probes against networks and computer systems requires discipline, resources, and above all, common sense and ingenuity. In most cases, there just aren't enough people and time to watch over systems or read the audit logs and trails

as there should be. Further, today's technology has developed and been deployed in a distributed, non-centralized manner. During this fantastic period of growth, many vulnerabilities associated with these new technologies are not understood in time to prevent incidents, or the response to such incidents is fragmented at best. In other cases, the protections are known but not implemented due to cost, resources, or lack of security motivation.

(4) **The chances of getting prosecuted are even lower, if even detected.** In order to prosecute, one needs acceptable legal precedent and plaintiffs. As to the laws, it will take some time for the lawmakers to understand technology and enact effective laws to combat computer crime. A key issue with the legal process is that the technologies and processes—in essence, the "speed" of things today—are so rapid that the law has a considerable amount of difficulty keeping pace with such changes. As for plaintiffs, large businesses such as banks and credit card companies are reluctant to prosecute cyber-thieves for fear that reporting a theft of funds will result in lost customers due to the public nature of court records in criminal lawsuits.

Information Warfare is not only being waged by hostile governments. Today's independent cyberthief can do the same—if not much more—damage to people and organizations with a keyboard than with a gun. As discussed above, financial institutions are not willing to divulge compromises and electronic thefts that occur with their systems for fear that their clientele will lose faith and take their business elsewhere. Industrial espionage via computer theft is very attractive as well. American companies spend billions on research and development only to have it electronically pilfered, thus leaving the victim company to pick up the research and development costs. And let us also not forget the true "cracker" who can gain a great deal of bravado by breaking into systems such as the CIA, Department of Justice, or New York Times web sites for the sheer notoriety of doing so. On this note, the number of such "hacks" and illegal modifications of prominent websites are being hailed by some as the new "gang wars" of the late 1990s with the media coverage and notoriety going to the person or organization with the "bigger and baddest" hack.

Information and information resources control or influence many aspects of the Western world's environment, culture, business, and way of life. As a result of "information warfare" becoming a Beltway catch-phrase, so too did the term "critical infrastructure" become a popular term given the reliance on certain key "systems of systems" that influence Western society. Certain national infrastructures are so vital that their incapacity or destruction would have a debilitating impact on the defense or economic security of the United States. These critical infrastructures include telecommunications, electrical power systems, gas and oil storage and transportation, banking and finance, transportation, water supply systems, emergency services (including medical, police, fire, and rescue), and continuity of government. Threats to these critical infrastructures fall into two categories: physical threats to tangible property ("physical threats"), and threats of electronic, radio-frequency, or computer-based attacks on the information or communications components that control critical infrastructures ("cyber threats").

This book is a blending of very old and very new concepts regarding the conduct of warfare. The old has been provided by Sun Tzu and the precepts contained in his classic *The Art of War,* a book commonly associated with the study of Information Warfare. The new has been provided by the recent advances in system technologies, communications, the Internet, and the new types of threats and attacks they bring our national security and business communities. Therefore, the purpose of *The Art of Information Warfare* is not to alarm but to educate readers to the problems arising from the world's growing reliance on information systems.

Thus, our goal in writing this book is to draw a parallel between the principles that governed the ancient forms of traditional warfare and the new forms of distributed, information-based warfare, applying them to both the corporate and national security environments in an objective, understandable manner.

The Commentators

The original *Art of War* includes a dialogue among several of Sun Tzu's top advisors, strategists, and generals. In keeping with this tradition, several portions of *The Art of Information Warfare* include dialogue among several of Sensei Sun and his most trusted and learned strategists:

- ***Sensei Sun***

  Sensei Sun was born in February, 1982, in the area known as Silicon Valley, USA. He was born to be a leader, and with his network interface, was far ahead of his time. Considered a prodigy, Sun was sending e-mail on the day he was born with his first words being, "The network is the computer." From the beginning, Sun spoke and understood the language of TCP/IP, which caused a close kinship with Yu Nix and the fledgling Internet. Sun, the brilliant strategist, can always be seen with a cup of Java in his hand and a perpetual sparc in his eye.

- ***Yu Nix***

  Yu Nix, born in 1969 at the Bell Laboratories, and was the result of breeding genetic algorithms with the goals of being simple and elegant, written in a high-level language rather than assembly language, and able to allow the re-use of his code. Yu Nix was born for a military career leading the forces of Information since his very heart and soul consisted of a small amount of assembly code referred to as a *kernel.* Yu Nix is a master of deception and has had several personas such as BSD and Linux to name a few. Indeed, Yu Nix has become the *de facto* leader of the Information Age, as his open language is the native tongue of the systems that drive the Internet.

Com Pak

Com Pak was born in Houston, Texas, in approximately 1981. Com Pak, unlike Sensei Sun, had a strategy to build smaller desktop systems for the village, instead of uniting the world. This strategy worked well, and Com Pak built the first highly-successful IBM-compatible computers. In fact, Com Pak attained Fortune 500 ranking and reached $1 billion in sales faster than any other U.S. company. However, Com Pak must also be feared. One must be wary of the small over the large, for is it not the minuscule virus, which cannot be seen with the human eye, that is able to render the giant helpless?

Sku Zi

Sku Zi began life around 1979 and was christened the Shugart Associates Systems Interface or "SASI." SASI was the first intelligent translator to interface between hard disks and minicomputers. When SASI reached adolescence, he changed his name to reflect his maturity, and Sku Zi became a prominent interface in modern personal computers and file servers across the computing environment.

Ji Wang, and Lan Wan are minor strategists in Sensei Sun's military machine.

PART 1

White Belt—Theory

Understanding the Basics

Overnight, a warrior is not made. The novice must learn patience and not act in haste, be humble not haughty, and have a keen, open mind. Only through a willingness to expand oneself beyond perceived limitations will the novice become worthy of the title Warrior.

As the novice progresses in his training he will come to bear the armor of Perseverance, the shield of Integrity, and the sword of Competence. Only then will this journeyman become a combatant on the cyber-field of battle.

It is through this and no other way that the novice becomes a Warrior.

Sensei Sun

CHAPTER 1

A Brief History of Cyber-Time

A thousand mile journey begins with one step.

Most of us have had the privilege of witnessing this phenomenon known as the Information Age. The maturation of technology has been staggering. However, one must always know where he came from, what his roots are and the strength of his foundation. This is true of all journeys including the trek through the Information Age.

It is interesting to track the evolution of Information Security (INFOSEC) as a function of the maturation process of technology. In the early days, most organizational computing took on the form of IBM or "Big Blue"—developed "Big Iron" systems. Large corporate, military, and government data centers sprouted up everywhere. And within this new environment, robust security staff, programs and products evolved to support the *centralized* computing model. Included in the centralized model were solid logical and physical access controls, backup and recovery mechanisms, and system logging tools. Networking was very limited by today's definitions and geographically restricted to direct or almost-direct connections to a mainframe computer system. Just when everything was secure and under control, along came a phenomenon known as client-server networking and the *Internet*. Rapidly, the centralized computing model became *decentralized* or *distributed* and the rules of the security game began to dramatically change.

The Internet realized an explosive, geometric growth pattern. In order to survive and compete in this new marketplace, organizations will need to conduct business and commerce over this new and exciting medium. While organizations are making the technological leap from the *centralized* to *decentralized* computing model they have *not* made the same philosophical leap in their approach to INFOSEC. In other words, they are still using billyclubs when they need to be carrying Uzis.

Historical Perspectives

Consider the analogy of earth history to our topic of the evolution of technology and INFOSEC. The "Big Iron" era can be compared to the prehistoric Paleozoic era or "old life". We are now nearing the end of a Mesozoic period or "middle life" with regard to the maturation of technology and the Internet. . . and on the cusp of a new and explosive era of technology, connectivity and commerce. How did we get to this point? The answer is simple—again, the Internet. The Internet is essentially, a network of networks, which began in 1969 as a gleam in the eye of the US Department of Defense as it searched for a fault-tolerant, redundant information network that could withstand the Cold War fear of a nuclear exchange. The Internet was then called ARPANet (Advanced Research Projects Agency Network) and was used to connect research facilities for collaboration in computer science projects. BBN (a company in Cambridge, MA) constructed the network and AT&T provided special communications lines to supplant existing phone lines, which were too slow and unreliable. ARPANet rapidly grew to more than 50 nodes and e-mail was the primary service used to augment project scientist's communication and productivity. However, membership in this exclusive "club" was limited to computer scientists with Department of Defense contracts. However, the desire for connectivity quickly spread to other parts of the academic research community.

While the Internet was evolving into the copper and glass highway that now crisscrosses the globe, UNIX became the *vehicle* of choice to drive on the highway. UNIX is an elegant, simple operating system invented by AT&T that became widely used in the academic world and has paved the way for today's international public communications system. In 1976, AT&T's Bell Laboratories created a UNIX utility called UUCP (UNIX to UNIX Copy Program). The UUCP utility was an efficient, low-cost way of passing files between computers via phone lines and designed to promote communication, fellowship, and give control at the user level.

Next, all that was needed was a *fuel source* to fill the tanks of the UNIX vehicles. That cheap, efficient fuel source was discovered in 1983, when ARPANet adopted TCP/IP (Transmission Control Protocol/Internet Protocol) as its standard for network communication. This protocol continues as the standard of the Internet today. ARPANet successfully demonstrated how a backbone infrastructure can serve as a connection between gateways to transfer messages between different networks. In 1986, ARPANet was phased out and replaced by the National Science Foundation Network (NSFNet), which became the US backbone for a global network. Opening up as the "Internet" for commercial networks and activities is a relatively recent phenomena which began as late as 1993.

Early Mesozoic INFOSEC

We have already established that most of today's cyber-cops are still using the billyclubs that may have been effective during the "Big Iron" era of INFOSEC. However, as the use of the Internet increased, so did the potential for outlaw behavior. Remember that *the Internet is a mirror of society with all the attendant goods and evils.* In November 1988, the first demonstration of a "cyber-evil" occurred when Robert T. Morris unleashed his "Internet Worm" program on the Internet. Morris' worm program replicated itself on thousands of connected computers in a matter of hours. To meet this new threat, the Computer Emergency Response Team was established at the Carnegie-Mellon University's Software Engineering Institute after the following press release:

> 12/13/88
>
> DARPA ESTABLISHES COMPUTER EMERGENCY RESPONSE TEAM
>
> The Defense Advanced Research Projects Agency (DARPA) announced today that it has established a Computer Emergency Response Team (CERT) to address computer security concerns of research users of the InterNet, which includes ARPANET. The Coordination Center for the CERT is located at the Software Engineering Institute (SEI), Carnegie Mellon University, Pittsburgh, PA.
>
> In providing direct service to the InterNet community, the CERT will focus on the special needs of the research community and serve as a prototype for similar operations in other computer communities. The National Computer Security Center and the National Institute of Standards and Technology will have a leading role in coordinating the creation of these emergency response activities.
>
> The CERT is intended to respond to computer security threats such as the recent self-replicating computer program ("computer virus") that invaded many defense and research computers.

The CERT will assist the research network communities in responding to emergency situations. It will have the capability to rapidly establish communications with experts working to solve the problems, with the affected computer users and with government authorities as appropriate. Specific responses will be taken in accordance with DARPA policies.

It will also serve as a focal point for the research community for identification and repair of security vulnerabilities, informal assessment of existing systems in the research community, improvement to emergency response capability, and user security awareness. An important element of this function is the development of a network of key points of contact, including technical experts, site managers, government action officers, industry contacts, executive-level decision makers and investigative agencies, where appropriate.

Because of the many network, computer, and systems architectures and their associated vulnerabilities, no single organization can be expected to maintain an in-house expertise to respond on its own to computer security threats, particularly those that arise in the research community. As with biological viruses, the solutions must come from an organized community response of experts. The role of the CERT Coordination Center at the SEI is to provide the supporting mechanisms and to coordinate the activities of experts in DARPA and associated communities.

The SEI has close ties to the Department of Defense, to defense and commercial industry, and to the research community. These ties place the SEI in a unique position to provide coordination support to the software experts in research laboratories and in industry who will be responding in emergencies and to the communities of potentially affected users.

The SEI is a federally-funded research and development center, operating under DARPA sponsorship with the Air Force Systems Command (Electronic Systems Division) serving as executive agent. Its goal is to accelerate the transition of software technology to defense systems. Computer security is primarily a software problem, and the presence of CERT at the SEI will enhance the technology transfer mission of the SEI in security-related areas.

Since the creation of the CERT, the following statistics have been compiled to illustrate the escalation in incidents reported. It is interesting to note that in 1997, an actual decline in incidents was realized.

Incidents Reported

Year	Incidents	Year	Incidents
1988	6	1993	1,334
1989	132	1994	2,340
1990	252	1995	2,412
1991	406	1996	2,573
1992	773	1997	2,134

Middle Mesozoic INFOSEC

With the creation of CERT, new tools and techniques were developed and made available to protect systems in the new distributed networked environment. These products were predominately developed for the UNIX environment and included a number of host-based tools such as TRIPWIRE, COPS and TCP Wrappers. These tools were developed to assist system administrators in their quest to secure their systems. However, these products were not easy for the uninitiated to implement and most required constant and tedious attention. Meanwhile, most INFOSEC departments continued their original charter, which was to protect mainframe resources, and thus remained unprepared to deal with the distributed, networked computing environment. Security implementations for host-based systems and organizational interfaces to the Internet were left to the discretion and motivation of network operations staff instead of the INFOSEC department. Concepts such as the secure configuration of Domain Name Servers (DNS), Internet mail, and packet-filtering routers added even more confusion to the Mesozoic INFOSEC department.

In the meantime, a hacker community was burgeoning. "The Masters of Deception" and "The Legion of Doom" became virtual clubhouses for the likes of Eric Bloodaxe, Phiber Optic, Prime Suspect, and Anthrax. These journeymen were extremely bright, self-taught individuals who passed on their secrets and tricks like Yoda to a Jedi Knight. No system was safe—from those owned by the Department of Defense to telecommunications giants such as NYNEX. The more exotic and high visibility the target organization was, the more bravado was achieved by hacking into and "owning" the target's networks and systems by obtaining root-level system access. In answer to the increasing cyber-threat, another breed of security product appeared to counter the threat—*firewalls*. These products were developed as choke points between the organization's systems and the public Internet. Firewalls started off as basic packet-filters and morphed into products with proxy capabilities and sophisticated logging capabilities. Again, the traditional INFOSEC department became even further confused with the new needs—to construct firewalls, electronic demilitarized zones (DMZ), and stronger proxies. The hackers had progressed to using gunpowder while the INFOSEC department continued to fight battles with rocks, clubs, and pitchforks.

Late Mesozoic INFOSEC

This is the period in the brief history of systems and INFOSEC in which we find ourselves today. As stated earlier, we are on the cusp of a technological explosion. We are also on the verge of a potential technological disaster—the "Millennium Bug" or Year 2000 (Y2K) issue. While a detailed discussion of the Y2K issue is beyond the scope of this book, a discussion of its potential impact is not. The good news is that many organizations are implementing new systems that will be Y2K-compliant and should not hiccup on New Year's Day 2000. The bad news is that a lot of mission critical applications are still running on old equipment and obsolete code that will cease to operate properly once the celestial odometer turns over from 1999 to 2000.

Some predict the Y2K issue may turn out to be as catastrophic of an event as the 1929 stock market crash with one major difference—Y2K will have a more profound impact on a wider audience—humanity as a whole. As such, the resources of most IT organizations are being consumed by the Y2K issue. As a result, security budgets are taking a backseat to fixing applications by reviewing and modifying date ranges on a line of code (LOC) basis. With estimates are that a line-by-line analysis of software costs about $1.00 per LOC, one can easily surmise that Y2K is not a cheaply resolved issue.

Another major event that will impact the modification of world computer systems in the immediate future is the transition of the European consortium from individual country currencies to the Euro-dollar standard. This also will have a tremendous impact on the IT departments that need to address both Y2K and Euro-dollar issues concurrently. One also wonders how Wall Street systems will handle "Dow 10,000" when the market index grows from four to five digits. Y2K, meet Down Ten K.

Many other issues are also converging on this mystical date. The growth of the Internet is expanding exponentially and businesses are expanding their capacities to conduct electronic commerce on it. Another example is the INFOSEC product industry. Mergers are daily occurrences with smaller companies being eaten and digested with the voracity of predatory jungle animals. What were a dozen reputable security product manufacturers has been consolidated to about three or four "mega-vendors" with varying levels of product quality, features, and reliability.

New INFOSEC products are being developed such as real-time intrusion detection systems (IDS) and cyberforensic tools that are making the detection, capture and analysis of digital evidence for prosecutorial purposes possible. Of particular interest is that the legislation necessary to convict hackers and other forms of technology abuse is finally maturing to the extent that cases can be tried and won in a relatively timely fashion.

The INFOSEC department has evolved to the early Mesozoic INFOSEC period while the cyber-criminals are using "Cyber-Uzis," "i/o hazardous" materials, and other weapons of mass cyber-destruction.

INFOSEC Countdown to Y2K

Several factors that will profoundly impact INFOSEC will come into play on December 31, 1999 and after, of course. These will include the geometric growth of Internet usage; a continued emphasis on the Y2K issue both before and after New Years 2000; and the enactment of stronger legislation to thwart cyber-crime while insuring individual privacy, and the maturation and convergence of INFOSEC technologies. Let us investigate each of these issues:

Exponential Growth of the Internet

A visit to ITM Solutions web site (http://itmsolutions.com/) displays some very interesting statistical data regarding the exponential growth of the Internet especially in the area of the world wide web. Some of the statistics are:

- Web site host devices in the U.S. increased 17% every 6 months from the time period July 1996 to July 1997. (Source: *Network Wizards,* 8/97)

- Argentina's web sites increased 50% every 6 months during the same time period and Venezuela's increased 94%. (Source: *Network Wizards,* 8/97)
- The *Wall Street Journal* reported that were 58 million users of the Internet in the U.S. and Canada as of 12/97. (Source: *Wall Street Journal,* 12/97)
- The U.S. tops the percentage of total worldwide Internet usage at 54.7% with Japan as the next top user at a distant 7.97%. (Source: *Internet Industry Almanac,* 1/12/98)
- According to IDC, in 1996 $2.6 USD worth of transactions were conducted via the world wide web. (Source: *IDC,* 7/22/97)

The statistics inarguably illustrate the fact that the Internet is growing at a phenomenal rate. Also poised to follow this growth is the parallel growth in commerce that will be conducted over this new and exciting medium.

Y2K Issues

As stated earlier, the Y2K issue is expected to have a profound impact on humanity. To make matters worse, not only is there a problem with the Y2K issue itself, but many applications have not been designed to handle the fact the *year 2000 is also a leap year*. Now tack these two items onto the Euro-dollar conversion issue discussed earlier and you have a full-blown IT melt down. Pundits are urging everyone within earshot to take drastic actions such as stockpile enough food, batteries, blankets, water and money (cash or precious metals) to last at the very least, until January 31, 2000.

What will happen as the result of these system failures? Some—but certainly not all—of the *possible* scenarios include:

- Banks will not be able to wire-transfer funds, access customer account information, and/or their ATMs may not work.
- Employees of corporations with electronic access keycards may not be able to enter or exit their workplaces or particular areas of such facilities like computer centers.

- Grocery stores will not be able to check out customer orders since scanners will no longer work, and/or customers may not be able to use ATM cards to pay for purchases.
- Utilities (power, water, natural gas) will no longer be available or guaranteed since their distribution is predicated on computer systems. Because of the layout of the national power grids, while one power company may be Y2K prepared, its links to other companies or grids (that may or may not be prepared) places it and subsequently its customers at risk of service disruptions. *Being "Y2K-compliant" will not guarantee "no problems."*
- Gasoline pumps will not operate, but that will not matter, since the computer modules in some automobiles may no longer function. The disruption of Galaxy-4 satellite service in early 1998 paralyzed gas stations across the country, and at Y2K, just as in 1998, customers may not be able to pay at the pump.
- Train and airplane travel will be impossible due to system malfunctions in the switching circuits on railroads, or the Global Positioning Systems may need to be reset.
- Some stock markets may crash and some businesses will fail.
- Although designed to survive a nuclear war, vast portions of the Internet may be unreachable (due to unprepared phone or power companies) while other segments will remain operational.
- Basic telephone and other forms of communication such as fax, cellular, and pagers may be disrupted or unreliable.
- Water will not be fit to drink since the computer systems that regulate purity by dispensing the proper levels of chemical agents or filtering processes will fail.

These are just a few of the *possible* scenarios, which do not paint a pretty picture. Large businesses (such as banks and brokerages) and infrastructure providers (power, water, among others) have not been publicly forthcoming in reporting their Y2K preparedness levels to their customers, choosing instead to place business interests ahead of customer peace of mind...but by doing so, they are placing customers at risk should their systems fail or be disrupted at Y2K. In the mad rush to deal with Y2K, panic overshadows logic. Thus, regardless of what is preached on corporate advertisements or the news, remember that ***Y2K-compliance does not guarantee no problems.***

Where will this all lead? As with the 1929 stock market crash analogy made earlier, we believe that significant changes will be made after the "great Y2K crash". For starters, there will be significant regulatory provisions enacted to protect the security, integrity, and availability of mission-critical systems. Government oversight with regard to computer security will increase substantially especially for financial and health-related institutions. On October 13, 1997, *The Report of the President's Commission on Critical Infrastructure Protection* was released. One of the recommendations is for, "The National Institute of Standards and Technology (NIST) and National Security Agency (NSA) work with the [then] proposed Office of National Infrastructure Assurance to offer their expertise and encourage owners and operators of the critical infrastructures to develop and adopt security-related standards." Now, everyone knows the speed with which the United States government operates. As such, the development and adoption of security-related standards will not happen prior to the havoc which will be wrought in the wee hours of January 1, 2000. That New Year's Day will mark the end of the Mesozoic INFOSEC era.

Legislation

The *Computer Fraud and Abuse Act of 1986* is the main body of legislation that has governance over most common computer crimes although other laws may be used to prosecute different types of computer crime. The act amended Title 18 United States Code 1030 and complemented the *Electronic Communications Privacy Act of 1986.* The privacy act outlawed the unauthorized interception of digital communications. The *Computer Abuse Amendments Act of 1994* expanded the 1986 Act to include the transmission of computer viruses and other malicious code. In addition to the federal laws of the United States, some states and foreign countries have passed legislation that defines and prohibits computer crime. As the first official information security team at the U.S. House of Representatives, we had several cases involving malicious activity that appeared to originate from foreign sources. We never got very far investigating these cases because the offending countries did not have legislation that would facilitate our investigations.

However, crimes are beginning to be tried under federal legislation. HR 1903, passed by the 105th Congress, entitled *The Computer Security Enhancement Act of 1997,* amended the *National Institute of Standards and Technology* Act to:

> enhance the ability of the National Institute of Standards and Technology to improve computer security, and for other purposes. The purposes of this Act are to (1) reinforce the role of the National Institute of Standards and Technology in ensuring the security of unclassified information in Federal computer systems; (2) promote technology solutions based on private sector offerings to protect the security of Federal computer systems; and (3) provide the assessment of the capabilities of information security products incorporating cryptography that are generally available outside the United States.

The two most important U.S. federal crime laws are 18 USC Sections 1029 and 1030. Section 1029 prohibits fraud and related activity that is made possible by counterfeit access devices such as Personal Identification Numbers (PINs), credit cards, account numbers, etc. However, in order to prosecute under 1029, the offense committed had to involve interstate or foreign commerce. Section 1030 prohibits unauthorized or fraudulent access to government computers and established penalties for such access. This section is one of the few pieces of legislation that deals solely with computers and also gives the FBI and Secret Service jurisdiction to investigate the offenses defined by the overall Act. The 1998 Presidential Decision Directive (PDD-63) that created the National Infrastructure Protection Center (NIPC) as the country's "IW" center is another example of new legislation regarding these uncharted waters of cyberspace.

The importance of legislation is obvious in defining and trying cases that deal with the abuses of computer systems. We would like to draw your attention to the less obvious advantage of legislation. Up until now, organizations were concerned with thwarting cyber-criminal activities. Now the emphasis will expand to include not only thwarting activities, but developing laws, standards, and acceptable methods of capturing the cyber-forensic evidence in *a manner that will stand up in court*. This concept will be important as we look at the changing role of the INFOSEC office.

Evolution of INFOSEC Tools

The early tools like RACF and ACF2 were created to protect access to files on the *centralized* mainframe. As the computer processing model matured to the *distributed* or network model, the tools that were developed remained philosophically centralized, host-based, and mainly UNIX-oriented:

Paleozoic Tools (Mainframe based)

Resource Access Control Facility (RACF) and Access Control Facility 2 (ACF2) are mainframe security utilities to restrict access to files.

Early Mesozoic Tools (UNIX host-based)

tripwire—written by Purdue University's Gene Spafford and Gene Kim checks for file integrity on host systems.

trimlog—written by David Curry to help manage log files on host systems.

tiger—written by Texas A&M's Doug Schales, is a set of scripts that scans a UNIX host system for security problems.

COPS (Computer Oracle and Password System)—written by Dan Farmer and Gene Spafford is a set of scripts that checks a UNIX host system for security holes.

tcpwrapper—written by Wietse Venema allows system administrators of UNIX systems to monitor and filter incoming network requests.

Middle Mesozoic Tools

The Middle Mesozoic era saw the development and deployment of tools to address specific needs such as *virus detection and eradication* software tools. This era also saw the beginnings of *encryption products* for common use and also during this period we saw the early forerunners of firewall technology. *Cyberforensic* tools and techniques were being developed for government and law enforcement use only. Lastly, another security technology that was beginning during this era was *network scanning* tools.

Late Mesozoic Tools

Firewalls—these technologies matured and became commonplace in organizations with Internet connectivity. (e.g., CheckPoint Firewall-1, Trusted Information Systems Gauntlet, etc.)

Encryption Products—Pretty Good Privacy (PGP) was made available to the Internet community at large by Phil Zimmerman, and has since evolved into the *de facto* Internet encryption standard. Other types of plug-and-play encryption products also became available.

Network Scanners—these technologies such as Internet Security Scanner matured and were used by system administrators to scan hosts or whole network segments (such as Class B or C segments) to assess host vulnerabilities.

Network Intrusion Detection Systems (NIDS)—we are now seeing the early stages of real-time intrusion detection systems (such as Real Secure and SHADOW), which are based somewhat on network scanner technology. Current NIDS solutions capture network traffic, analyze the traffic for intrusive signatures, log the data for evidence, and potentially take an action based on the severity level of the intrusive signature such as paging system administrators, shutting systems down, and so forth.)

Virtual Private Networks (VPNs)—are another technology that is in the early stages of development. The concept of a VPN is that one client can communicate with another client over an untrusted network (e.g., the Internet) by establishing an encrypted pipeline between the two clients that lasts for the duration of the communication session.

Certificate Authorities (CAs)—are a technology that is somewhat a derivative of encryption technology. CA is a hot technology since it will support electronic commerce over the Internet. The concept behind CA is that in the process of conducting an electronic transaction, a trusted third party (i.e., the CA) will verify to the business that the customer is who he claims to be and vice versa.

Summary—INFOSEC Meets Corporate Security.

We are finally at the end of our journey through the Paleozoic to the Late Mesozoic periods in the evolution of technology, the Internet, and the INFOSEC office. As we have attempted to explain, the INFOSEC department has lagged behind the cyber-criminal element philosophically, legally, and without proper defenses. Our view of the transition that will take place in the future INFOSEC office is similar to that of an English Bobby whistling and swinging his billyclub as he walks along a cobbled street to a member of a police SWAT team in a crisp black uniform with body armor rappelling down the side of a building with an automatic weapon slung across his back. The distinction is that the future INFOSEC office will function with stealth and speed and will be well armed both legally and with the proper weaponry. Some other salient points of the new Security Office include:

The INFOSEC function will merge with the traditional Corporate Security office. This will occur because traditional physical security functions are now and will continue to become systematized. There will be little difference between physical and logical break-ins. Both will be investigated, forensic evidence captured and analyzed, and perpetrators will be prosecuted using the same or newer, integrated subsystems. Networks and the systems that reside on them will be monitored in real-time as surveillance cameras are currently monitored. Computers and their resident information have become the new "cores" of the corporation—all available security resources are now supporting the protection of information and proprietary data through traditional computer security, personnel background investigations, photo ID cards, and alarmed frontiers. Indeed, the two functions are merging into a consolidated security office.

INFOSEC will function in real-time. Past and some current security philosophies rely on logs, which alert INFOSEC and system administrators after the fact in a reactive mode. Network intrusion detection systems will monitor networks in real time and may be based on neural network technologies so that network characteristics and intrusive behavior can be "learned" just as the human immune system "learns" when something is wrong with the body. Security staff will take action as events occur in real time and shut systems down, redirect traffic, or "spoof" intruders with false replies, in a manner as swift as a police SWAT team. Thus, security teams will be more proactive and less reactive.

Security will morph into an incident response capability. The tools will be there to react to incidents in a real-time manner, the tools will support the capture of evidence that will stand up in court, and the laws will be available to try the case. The new security office will have policies and procedures in place to: (1) categorize levels of incidents, (2) the expected responses to incidents, (3) escalation procedures, (4) interfaces with law enforcement, (5) the press, (6) containment procedures, and (7) postmortem analysis of incidents to minimize or preclude future occurrences. Security staff will be trained and certified in all aspects of incident handling procedures.

It is interesting to note that some organizations—private and public—have already evolved into this new Security Office model, with a very high success rate in conducting their corporate security duties.

Now let us bask in the light of Sensei Sun's wisdom

CHAPTER 2

Targets and Vulnerabilities

Let them first climb on the roof and then take away the ladder.

This ageless axiom advises the warrior to bait his adversary into a situation where there is "no way out." In preparing for Information Warfare, one must fortify his castle with proactive layers of security, thereby creating his own defensive paths and direct the defense instead of following the dictates of the attacker.

On ne passe pas—"they shall not pass", was engraved in the Maginot line, a military fortification designed in the early 1900s to prevent a future German invasion of France. Karl von Clausewitz, the brilliant Prussian military theoretician and instructor, wrote a hundred years earlier, "If you entrench yourself behind strong fortifications, you compel the enemy to seek a solution elsewhere." On 10 May 1940, General Erich von Manstein, a student of Clausewitz's writings, demurred at the thought of attacking the well-fortified line and slipped his German armor through the Ardennes forest.

The Ardennes was considered a poor place to deploy armor and without the Maginot Line it would have been the worst choice. But the strength of the line changed the dynamics of the situation and made the previously impenetrable Ardennes look like the most workable solution—underlining the veracity of Clausewitz's observation. And because the French had no strategic reserve to shield themselves from an attack from that direction they lost their territorial sovereignty in just ten days.

In the south where the Italians had no choice but to attack the line, seven French soldiers operating behind the controversial fortification, held up an entire enemy division for more than a week.

Wouldn't Maginot be an appropriate name for a firewall product or any company offering *only* technical solutions to communications security problems? Good firewalls and other purely technical solutions do their work effectively, but to a clever and determined attacker they are just obstacles to be either broken or side-slipped, whichever is most effective.

It is not just the financially motivated cyber-thief or teenage hacker that is testing the electronic Maginot lines of global corporations. Terrorists and states unsatisfied with the current balance of power are turning to what they consider to be low-risk, high-return cyber-strategies that avoid traditional types of military defense. According to George Tenet, Director of the Central Intelligence Agency and statute head of the United States intelligence community, "It is clear that nations developing these programs recognize the value of attacking a country's computer systems both on the battlefield and in the civilian arena." He pointed to telecommunications and banking as high-profile targets.

Technology, combined with the creative genius of military thinkers around the world, is leading to the development and application of new forms of warfare, and the innovative modification of traditional military practices. While the United States and its allies are the source of much of this innovation, others are motivated by the dominant military position of the United States, and its demonstrated commitment to maintaining its military lead. This basic reality is forcing many of the nation's adversaries (current and potential) to seek other means to attack American interests. Lieutenant General Patrick Hughes, USA, Director of the Defense Intelligence Agency in Washington, DC, mentioned these (and several other) items in a recent Congressional testimony. With regard to this book, some of the more important vulnerabilities and opportunities are listed below:

Information Warfare (IW) involves actions taken to degrade or manipulate an enemy's information systems while actively defending one's own. Over the next two decades, the threat to American information systems will increase as a number of foreign states and sub-national entities emphasize offensive and defensive information warfare strategies, doctrine, and capabilities. Current information on national vulnerabilities, and foreign intelligence initiatives in general, point to the following threats:

- Trusted insiders who use their direct access to destroy or manipulate the information or communications system from within.
- Modification of equipment during transport or storage.
- Physical attack of key systems or nodes, including the insertion of modified or altered hardware.
- Network penetration to include hacking, exploitation, data manipulation, or the insertion of various forms of malicious code.
- Electronic attack of various interconnecting links, sensors that provide data to the system, or other system components.
- Empowered agents including "sponsored" or individual hackers, cyber-terrorists, criminals, or other individuals who degrade, destroy, or otherwise corrupt the system. In the most advanced case, empowered robotic agents, embedded in the system, could be used to take autonomous (timed) actions against the host or remote systems or networks (cyber war).

Cybernetic warfare (CYW) is a distinct form of information warfare involving operations to disrupt, deny, corrupt, or destroy information resident in computers and computer networks. One particularly troubling form of "war in cyberspace" is the covert modification of an adversary's data and information systems. This form of warfare will grow in importance as technology makes new methods of attack possible. Cybernetic warfare defies traditional rules of time and distance, speed and tempo, and the conventional or traditional military capabilities of the opposing elements.

Transnational Infrastructure Warfare (TIW) involves attacking a nation's or sub-national entity's key industries and utilities—to include telecommunications, banking and finance, transportation, water, government operations, emergency services, energy and power, and manufacturing. These industries normally have key linkages and interdependencies, which could significantly increase the impact of an attack on a single component. Threats to critical infrastructure include those from nation-states, state-sponsored sub-national groups, international and domestic terrorists, criminal elements, computer hackers, and insiders.

Asymmetric warfare—attacking an adversary's weaknesses with unexpected or innovative means while avoiding his strengths—is as old as warfare itself. In the modern era, many forms of asymmetric attack are possible—to include the forms of warfare outlined above, terrorism, guerrilla operations, and the use of Weapons of Mass Destruction (WMD.) As a result of the dominant American military position on the world stage, it is very likely to be the focus of numerous asymmetric strategies as weaker adversaries attempt to advance their interests while avoiding a direct engagement with the United States on its own terms. If forced into a direct conflict with the United States, those same adversaries are likely to seek ways of "leveling the playing field" to maximize their chances of success.

Asynchronous warfare involves a preselected, or delayed (timed) attack on an adversary, taking advantage of the passage of time to develop a strategic opportunity or to exploit a future vulnerability. In a preselected attack, the operation has a latent effect on the adversary. Human or technical assets are strategically placed well before—sometimes years before—the actual confrontation. In a delayed attack—often carried out as an act of reprisal months or even years later—the operation is conducted after an opponent has lowered his guard.

Essentially, in the Age of Information Warfare, one is either a target or a victim. In other words, a target has defenses against attackers while victims are defenseless. On a national, strategic level, there are a number of intriguing target possibilities, including:

Electronic Switching System (ESS)—Nationwide system that manages all telephone communications. Consider the consequences if the nation could not communicate via the telephone or dial-up Internet access.

Global Positioning System (GPS)—US-developed constellation of geosynchronous satellites that provide excellent navigational data for civilian aircraft, ships, and handheld units used by campers. Provides *precise* information to US military units and attack systems.

Internet—the communications backbone of science, industry, and society.

Commercial Operating Systems and Applications—This is an accident waiting to happen. What about commercial off-the shelf operating systems that run major networks for large government agencies and companies? Who knows what lives inside these "untrusted binaries" in such widespread use around the world? Users and administrators must be on constant alert to the almost-weekly announcements of a new vulnerability in these systems and be prepared to implement corrective action immediately to avoid potential threats to the integrity of their data and networks. The same can be said for financial and other business-critical applications that are used in conjunction with these untrusted operating systems.

One serious vulnerability not discussed in many circles is the sad but true fact that the mission-critical systems and infrastructures (financial, power, and most business or government systems) of the United States and elsewhere are run by commercial operating systems and software applications purchased *with the assumption that such products are secure as shipped from the manufacturer.* Unfortunately, this is not the case, and numerous vulnerabilities have been discovered in systems that were marketed as allegedly "secure" to industry or government specifications. Why? Some software companies are more concerned with profit, market share, and putting competitors out of business than they are with producing a quality software product that provides reasonable levels of security and acceptable levels of risk to the user. Granted, total security is as real as the Tooth Fairy, but stronger quality assurance must be taken on these products the world is now relying on.

Today, unfortunately, slip-shod products are rushed to market quickly, being driven by their competitor's schedule or their own internal marketing efforts. This effectively turns the consumer and corporate markets into expanded, "beta" testers who, instead of *being paid* to examine a piece of software for quality, *pay* the manufacturer for the privilege to own a license for an untested product and stand a good chance of having to absorb the costs of securing, recovering, or restoring their systems and data resulting from issues arising from a shoddy product nobody outside of the vendor has examined! During their use or "examination" of such products, systems routinely crash, data gets lost, or other issues arise that comes from implementing such untested software. While not an "external" attack to information systems like a hacker or cracker, such untested software applications are an equal threat to the sanctity of corporate data and information resources and the infrastructures relying on such products.

It was rumored that the Microsoft Windows 95 operating system installed by the consumer masses, shipped with over " 5,000 *known* bugs." Not unexpectedly, corporate and consumer clients complained about the quality of the new operating system when it shipped amongst much fanfare in August 1995. After several "service fixes" to the product, an upgraded operating system, Windows 98, was rushed to market in late June 1998 in spite of an ongoing United States government court case, and reportedly fixed "about 3,500 *known* bugs in Windows 95." Ironically, a service pack, euphemistically dubbed a "Multimedia Enhancement Pack" was released less than sixty days after this product hit the market to not only enhance the product, but "quietly fix" some of the bugs that were still in the product when it shipped! *A quality product?* Sure, if the company considers the world consumers to be unpaid "quality assurance" or "continuing beta testers" for such software. In the same vein, the auto industry recalls vehicles with defects in them and fixes such defects at no charge to the "owner"...but the software industry requires that its "users" / "owners" fork over money to get such defects and dangers to their data fixed. However, as of January 1999, even with a "service pack" and a few "patches" to the operating system, Windows 98 is still not Y2K-compliant. At this late date, a quality product must—by definition—not fail on 1 January 2000. To be fair, the Windows 98 product is more stable and robust than its predecessor, although it ships with several controversial features seemingly placed in the product for product placement than end-user utility.

Of course, any product or operating system requiring a one hundred megabyte "service pack X" should raise an eyebrow or two...Most other reputable vendors, when releasing such large updates to a product, assign it a new version number (e.g., 2.6, 2.7, etc.) and free the end-users and systems administrators from nomenclature confusion over whether the update will transform their product into an OEM, VAR, SP-1, or SP-2 "beta 3" release version with unknown new dangers to their data and information.

The installation of and subsequent reliance on such systems that have not undergone peer review or independent analysis is an accident waiting to happen. While items like UNIX (an open operating system that "runs" most of the Internet), Pretty Good Privacy (the *de facto* Internet encryption tool), and Netscape Navigator (the first, and some would say *only* reputable Web browser) have released their source code to the world for public analysis, disclosure, and discussion, many of the world's largest operating system and applications vendors—particularly Microsoft—do not, citing "proprietary trade secrets." In these cases where software has undergone worldwide peer review, the result is that user concerns and quality control issues are addressed *before* the product hits the open market, not after, where a considerable user base exists and is potentially threatened by bad code. Further, users have the opportunity to see how the programming code will interact with existing applications, much like checking a medical prescription for any potential drug interactions or side effects. Software that has been examined by "independent third parties" stand a better chance of being accepted as indeed "secure" and "stable" than products where the vendors announce "our product is secure...trust us!" In this case, an objective, third-party "Software Underwriters Laboratory" for instance, would not be a bad thing.

An example of the user community's reluctance to sleep well and rely on untested proprietary software is found in government circles in the early 1990s when the National Security Agency and National Institute of Standards and Technology attempted to create a standard encryption system for the United States to replace the antiquated Data Encryption System (DES). "Use it," they said in official reports, "but the encryption algorithm is classified TOP SECRET and not available for independent review." While the implication was "trust us, we're the government"—the product flopped and was declassified in mid-1998. Some would argue that the reason why UNIX, PGP, and Navigator became *de facto* user products in the computing community was that the software was reviewed by outside experts who certified the products, algorithms, or software code were robust, stable, and worked as advertised or intended.

While lucrative for security professionals, the increase in known vulnerabilities associated with such "proprietary" systems is disheartening. Where is the product security, stability, and reliability for the "good of the customer base"? Recently, under tremendous pressure from the deep-pocketed software industry, new copyright laws passed in 1998 prohibiting reverse-engineering and analysis of computer software without the express consent of the vendor. After much wrangling from the security community, Congress finally inserted provisions for academics and security professionals to be legally able to analyze software for security or academic purposes only. If the industry continues to develop insecure, untested, programs and operating systems—and prohibits independent testing and analysis—the future for truly secure operating systems—and systems in general—is fading rapidly from reality.

In July 1998, news surfaced that the Navy's first Commercial-Off-The-Shelf ship, the Aegis vessel USS Yorktown, had a systems failure only hours after departing Norfolk. The ship's Windows NT network crashed and rendered the vessel unable to continue its mission. Why? Who knows what other applications interacted with the NT software to cause the crash. Can the Navy dissect the NT operating system to find the flaw like they can in UNIX? Not a chance. The hacker and quality assurance communities had a field day with this latest blunder, dubbing Windows NT as "Needs Towing" and an operating system that certainly "Needs Tweaking." Yet the Navy is going ahead with plans to standardize fleet information systems to this allegedly "secure, stable, and robust" operating system, most certainly out of user familiarity with its interface that is nearly identical to many home computers. Sadly, the Marine Corps is following the Navy's IT-21 Project and standardizing the Corps on Windows NT as well as most of the Department of Defense and United States government.

One final note. The majority of commercial software and services are produced by American companies, many of which are written by foreign nationals employed by the software companies working on visas in America or back in their home nations. This is a major concern to government organizations who try to monitor personnel with critical access to systems and information. How easy might it be to co-opt a programmer in India to place some small backdoors for the Indian government to have secret access to any Windows NT server? Given the poor quality assurance measures in the industry today, our guess is very easy. Suppose these Indian programmers inserted some malicious lines of code into NT as a way of "getting back" at the US after it imposed economic sanctions on their country after their recent rounds of nuclear testing in early 1998? Not a pleasant thought, but a very real vulnerability. A good number of programmers and consultants working the Year 2000 issue are foreigners who are granted nearly unlimited, unfettered, and unmonitored access to the mission critical systems of our largest corporations and government organizations without any criminal background checks. Need we say more?

There are hidden programs, routines, and "Easter eggs" such as small flight simulators and pinball games hidden inside such products by their programmers, which perhaps adds to the size, complexity, and problems running the software. *Do we really need to play a flight simulator in a spreadsheet? Will the Navy or the rest of the world know what evil or "treats" lie in the 40-million-plus lines of programming code that constitutes Windows NT, Internet Explorer, or Word?* Probably not. *Will we still run the software and put up with the crashes, hiccups, and reboots associated with these products?* Sure...it's a "feature" and a seemingly acceptable level of risk to the world. Unless the NT server crashes and the famed Blue Screen of Death appears while targeting a Harpoon missile, that is.

Sadly, most policymakers, flag officers, and corporate executives are not products of the Communications Revolution. They do not understand programming code, the critical value of information, or the inherently "virtual" way the world works, not to mention the vulnerabilities inherent with the growing reliance on information infrastructures. Everyone plans for the major military offensive through the procurement of high-profile and glitzy weapon systems, but no one is planning for the critical defense of our less visible—but equally critical—interior vulnerabilities, the "Soft Underbelly" of the country.

While a great deal of press attention has been focussed on the teenage hacker and the egomaniacal programmer gone wrong, these are actually the least threatening intruders as their motives are childish. The acts of these people can range from bravado to destruction, but they are most often aimed at getting attention or simple greed.

Terrorists and state-sponsored programmers are less likely to want attention guaranteed to stimulate defenses. They prefer to attach themselves like parasitic organisms to government and corporate systems either to create wider security breaches or simply create long-term taps into strategic information. This style of attack can be more insidious than a destructive attack, as stolen or corrupted information (which should be backed up anyway) never actually disappears from its owner. In human terms, each day the victim gets sicker, but never knows why until it is too late.

It does not take a genius to develop tools or applications to effectively bring down one of today's mission critical, commercial-off-the-shelf systems. Indeed, there are numerous free "hacker tools" and several legitimate diagnostic tools that can be used for both good and evil. In short, **the greatest vulnerability is uncertainty regarding the content and integrity of programs and operating systems that drive our commerce and protect our national security and corporate secrets.**

While some may scoff at the likelihood of large-scale attacks on corporate and government infrastructure through the medium of commercial software, remember how the best military experts prior to World War II considered the Ardennes to be an impractical axis of attack. In the security business, the very act of dismissing the possibilities of an attack raises the chances of its ultimate success. Without knowing the enemy's activities and routes into the fortress, the inherent risks to one's organization are present. Ignoring it will not make it go away.

The attack will come. A strong defense will be necessary.

Comparisons—Old and New

In crafting this comparison between the classic *Art of War* and the new *Art of Information Warfare,* we would like to make an even deeper analogy. This analogy will further compare the various elements of information security with the Japanese concept of a samurai. The ancient samurai had in his possession three items: (1) a sword, (2) a mirror, and (3) a jewel. These items were signs of royalty and can be translated into the concepts of power, health, and wealth. We will use these items to build our model information security infrastructure. The *katana,* or Japanese sword, will depict the power aspect of the discussion, the mirror will be used to demonstrate the art of introspection or health of the network infrastructure for defensive purposes and lastly, the jewel will represent the wealth of knowledge and continuous training.

The Japanese sword is considered to be unsurpassed when compared to any other form of iron-crafted art. Three timeless attributes apply to the crafting of the Japanese sword: *flexibility, rigidity, and cutting power.* Flexibility requires that the constituent metal be soft while the cutting power found on the edge requires that the metal be hard. To achieve both conflicting attributes within the one sword seems to present the swordsmith with an impossible task and interesting paradox.

To achieve this seemingly impossible task, one must understand the processes by which the sword is constructed. The process involves the layering of metal that is hammered and folded over again and again. The sword's inner core or *shingane* is made of softer metal with a lesser amount of carbon content than the outer layers. The soft inner core of iron gives the sword its flexibility. The *shingane* is then wrapped with the hard outer layer of iron, or *hadagane,* that is of a higher carbon content and gives the sword its cutting power and edge. The tempering process involves covering the entire blade with a clay and charcoal ash compound. The clay is scraped away prior to heating and the varying layers of thickness of the remaining clay causes the blade to cool at different rates. This causes the creation of a variety of crystalline compounds of iron and carbon along the blade's surface. At long last, after much patience and ingenuity, this layering and "welding" process produces the ultimate weapon. It is well-known that each master swordsmith has his own secrets and techniques in sword making, secrets that were passed down through the generations and were evident in the features and/or design of his weapon.

Our security model should be the same as the *katana.* The INFOSEC program must be flexible, powerful, and always have an edge. If your campus network is connected to the Internet, it must have a hard outer layer as is provided by a firewall, access control cards, and a strong physical security program. The inner core of the network's security model must be flexible for the users but resilient at the same time. This is the concept of providing host, or fileserver security for the devices that reside within the protection of the firewall.

Also, in building an INFOSEC program, consider the layering process of the swordmaker. Similarly, the program should also be layered to give it strength and flexibility. Layers include but are not limited to the aforementioned outer layer of a firewall or guard force, but employee education and awareness programs, antiviral programs, host security, and administrative security. As we traverse the journey of this book, we will make similar analogies between other parts of the *katana* and security implications. Remember that each component of an INFOSEC infrastructure such as routers, firewalls, and software tools changes on a regular basis.

Also remember that information is a double-edged sword. The knowledge on how to both develop or beat each type of countermeasure is widely available to both you and the adversary in a variety of mediums.

CHAPTER 3

Strategic Assessments

Flies never visit an egg that has no crack.

There are many methods to protect your organization's presence on the Internet, from screening routers to proxy firewalls. The stratagem here is that if there is a crack in your perimeter protections (i.e., the egg's shell) which exposes the inside, flies (hackers and crackers) will be attracted. Therefore, no crack; no flies. Likewise, hackers will not usually waste time on networks that do not have readily apparent vulnerabilities or cracks, as they wish to maintain stealth and not risk detection by repeated, overt probing. For example, a major vulnerability for a site is the modem. Ensure that you are using secure modems or other methods of secure remote access.

Sensei Sun

The tools necessary to conduct Information Warfare are available to anyone with a motive and a modem. The motives of conducting Information Warfare are the same as they are for conventional warfare—greed, power, politics, fear, and survival.

Lan Wan

Information is important to the citizen;
Information is essential to the state;
Information is the lifeblood of the nation.

Ji Wang

In order for information to have any value, it must be timely, accurate, and relevant.

Com Pak

Wars of the future will be fought with lines of code.

Yu Nix

He who has better code will win the skirmish,
He who has better algorithms will win the fight,
He who has the better information will win the battle,
He who has more knowledge will win the war.

Lan Wan

First there is data;
The accumulation of data becomes information;
Information begets knowledge.

Sku Zi

Knowledge must be guarded,
Information must be protected,
Data must be encrypted.

Yu Nix

Data will inflict a scratch;
Information can cause harm;
Knowledge will kill.

Sensei Sun

The movement of data is like the flow of water; it always seeks its own level.
The movement of data is like electricity, it will always seek the path of least resistance.

Ji Wang

By this, Sensei Sun means that data will naturally seek out those who will use it best.

Com Pak

No, Ji Wang, by this admonition, Sensei Sun means that once data has been set into motion, it travels with an inertia that may not be anticipated. Data, in its flight from branch to branch, may land in a tree where you do not want it to roost.

Lan Wan

Data can also hide and be given renewed life even if you thought it was destroyed. Much care and caution should be exercised when eliminating the media (i.e., diskettes, hard disks, etc.) on which the data was once resident. This media can be manipulated such that the "destroyed" data is resurrected for all to see.

Sku Zi

Before disposing of media-

In general, media containing regular data should be overwritten 3 times prior to destruction, media containing sensitive data should be overwritten 6 times, and media containing confidential data should be overwritten 10 times.

Sensei Sun

Information has an inherent duality—a Yin and a Yang, a dark and a light side, fullness and emptiness, solid and liquid. It resides at the crossroads between data and knowledge. Information is the mune of the samurai's sword, or the backbone that resides between data and knowledge.

Yu Nix

If you can protect a byte of data;
You can protect a kilobyte.
If you can protect a kilobyte of data;
You can also protect a megabyte.
If you can protect a megabyte of data;

You can also protect a gigabyte.
If you can protect a gigabyte of data;
You can also protect a terabyte.
All manner of protection are the same.

Sensei Sun

Data and information must be categorized.

Does not the ruler separate his gold, silver, and bronze into three piles? Does not the ruler have his scribes account as to how much gold, silver, and bronze is in each pile?

Gold is maintained under lock and key, silver is displayed throughout the house, and bronze is used for ordinary utensils and implements. In the same manner, data must be categorized, separated and protected as appropriate. Once categorized, data must be labeled in order to assess how much protection it needs, who needs to have access to it, and how to measure in the event that it is lost.

Yu Nix

All data that pertains to money is sensitive;
All data that is associated with people is private;
All data that describes the inner-workings and future strategy of the military, country or state is classified.

Lan Wan

Data, unlike gold, can be duplicated and stored in different locations—at the same time. A copy of the same data can reside in two places at the same time. Copies of the same data can reside in 1000 places at the same time.

CHAPTER 4

Preparing for Siege

Await the exhausted enemy at your ease.

In ancient forms of warfare, this stratagem had merit by letting the enemy march to your doorstep. By the time he gets to you, he has expended a great deal of effort, supplies, and energy in the journey. He will be easier to defeat than if he were fully rested. In the current arena of Information Warfare, the same tactic applies. The basis of this stratagem is to make things difficult for the hacker. Let them do all the work and make sure that you have no crack in your egg. He may get weary of trying to break into your systems and move on to an easier target.

Infrastructure Attacks: Strategic Considerations of Information Operations

> *"While ultimately military in nature, Information-based Warfare is also waged in political, economic, and social arenas and is applicable over the entire national security continuum from peace to war and from 'tooth to tail."*
>
> **National Defense University**
> **(Excerpt of 1997 Definition)**

The ongoing threat lies in the strategic implications of Information Warfare. Take, for example, the conduct of strategic bombing operations against Germany's industrial complex during World War II which undermined their war effort at home, forcing its government leaders to contend with civil unrest, failing industry, and food shortages in addition to their already complicated war efforts against the Allies on two fronts. Likewise, the German V-series of "terror bombs" showered random destruction on the innocent civilians of England. Later, in the Vietnam conflict, the Vietcong used hidden agents to wreak social havoc and create social unrest in the South Vietnamese civilian population. Since these "plants" could not be easily distinguished from the natives and could appear at any time, this hidden threat remained an effective terror weapon during the war. In the continuing saga of Middle East relations during the 1980s, hijackings and aircraft bombings prompted the civilian populace of the West to pressure the government for action and defense against these indiscriminate acts of violence or war. The 1995 Oklahoma City bombing demonstrated yet another terrorist act to cause civil unrest in the heartland of America. In today's information society, a logic bomb placed on a key information-processing system is yet another effective method to cause civil unrest and function as an Information Age terror weapon.

Clausewitz's Trinity of War describes the role of the People—civilians and society at large—in the overall national security effort. With civil unrest, a government is unable to devote full attention to external activities (e.g., military actions, diplomatic or economic activities) as it must also contend with an internal and quite angry civilian populace closer to the national leadership. As stated, computers, information, and the speed by which it is communicated in that void known as cyberspace drives today's society. This new environment encompasses the military and civilian command, control, and communications (C^3) infrastructure such as phone and television services, traffic signals, subway systems, financial institutions, and government communications-all run by computers in a semi-autonomous manner. However, this *Electronic Maginot Line* of internetworked systems that provides critical strategic services to the nation will grind to a halt if a key electronic artery is cut or attacked. Jamming or disrupting one of the dozens of communication satellites that supports the world information infrastructure is one example of such activity, as evidenced in the 1998 Galaxy-4 satellite incident that caused electronic chaos lasting over a week for much of North America. Along the many roads in America, one can see a sign that reads "CAUTION-BURIED CABLE." To a hostile information warrior, this sign is a welcome sight that, to him, reads "BURIED CABLE-CUT HERE."

Renowned author Tom Clancy's book series included a novel about just such an event. In *Debt of Honor,* a computer virus was released into the financial network of America with disastrous results to the stock market and global economics. While fictitious, such a situation could be created by a foreign nation, organized crime group, or terrorist organization as a tool for blackmail or a prelude for a larger, more widespread aggression. In a 1993 lecture in France, a well-known advocate of the utility of the Information Age, remarked to a collection of French military leaders that:

> I can destroy any major nation in 24 hours with one platoon of knowledge warriors and make billions of dollars on the international market by doing so, because I will know when this is going to happen and invest accordingly.

Winn Schwartau is praised with bringing a formerly-classified topic—Information Warfare—to public and national light and attention. His detailed discussions about such attacks and the world's inability to defend against them bring shocked looks to his audiences' faces. His testimonies on Capital Hill and elsewhere have been met with equal disbelief when he sadly, but correctly, observes that today's world seems to dictate that "the computer is right" and that what ever information the computer represents-credit reports, school or police records, and financial portfolios-is taken as fact. Imagine the chaos caused by a hostile organization conducting this type of offensive information operation on a broad scale. Such has been dramatized in recent movies and books, but this is a real threat and does not happen only on the battlefield between opposing military units, but also between one or more private individuals from anywhere in the world.

Crashing the stock market is a favorite strategic "what-if" by amateurs attempting to describe the potential of strategic information warfare (SIW), and will not be discussed here because the global implications of a breakdown in financial information systems are fairly well-known...all one has to do is read current news headlines to know the fragile nature of the global financial system. Further, a public misperception is that information warfare is limited to computers, and conducted only by computers, keyboards, and modems. We propose the following hypothetical situations demonstrating five SIW attacks, some of which are not solely computer-based:

Train Derailments or Air Traffic Control Mishaps: In February 1996, a commuter train collided with an Amtrak passenger train just outside of Washington, D.C. The current analysis points to faulty signals that gave a "false proceed" message to the conductor of the commuter train, who proceeded on the assumption that the signal was correct and he was clear to speed up. This particular railroad signal system was controlled from a nationwide control center in Tampa, Florida through the use of high-speed computer networks. The result of this faulty signal was twelve passengers killed by fire, many more injured, the repair costs for the trains involved in the collision, and the time spent on the federal investigation of the crash and signal system. Likewise, western air traffic control systems are computer-controlled to relieve the work burden of human controllers. While not completely internetworked, they are a key component of the domestic information infrastructure. *We're scared enough when something breaks down under normal circumstances. What if such computer-control systems were infiltrated, destroyed, or changed in an intentional nationwide information attack?*

Traffic Signals and Power Grids: In the latter part of 1995, a school bus in Illinois was damaged by a speeding freight train and several elementary school students were killed. This time, the cause was a faulty traffic signal on the road crossing the railroad tracks, which the school bus driver naturally followed. *We use the roadways to commute and travel. Millions of innocent people are exposed and are potential targets to an attack on local traffic- or subway-control networks. Blizzards, hurricanes, and tornadoes cause power outages each year, and we all know how annoying, troublesome, and in some cases, life-threatening they can be.* An intentional attack (or unintentional accident) could bring down power grids and paralyze a city. *Simply bomb the power plants and a few substations,* *and you've effectively shut down a city, state, or region—along with its computers, most communications mediums, and a majority of economic resources; most shops, businesses, corporations, and government buildings would close. Such a circumstance is what we refer to as "Functional Paralysis," or a sizable Denial of Service.*

Wag The Dog: We read the paper and watch the news daily. Through the use of a personal computer and some common commercial software, an information warrior can electronically edit or morph pictures and video. Imagine the many implications if Saddam Hussein were portrayed on Iraqi television sitting down to a state dinner with a glass of wine, roast ham, and buxom women surrounding him, when such actions are forbidden by his Islamic religion. *All it takes is video editing equipment to receive the actual signal and broadcast the doctored one on the same frequency in real-time. Our mothers always told us to "believe half of what you see, and absolutely nothing that you hear."* This is sound advice for a society that can alter reality *(or given the media-driven world today, the perception of reality)* with a keyboard and mouse. Edited pictures of the President, First Lady, and other public figures in compromising situations have already circulated around the world on the Internet, as have doctored copies of legitimate web pages. *How many edited pictures have appeared on the front page of leading Western newspapers already-and been subsequently accepted as "fact"?* This is a perfect example of classic military deception, psychological operations, and perception management. By altering the perception of reality, one can destroy an enemy's will to fight, support for national leadership, or win the "hearts and minds" of his adversary. Later chapters will discuss this type of "information warfare" directed at the actual heart, mind, and soul of the human being. In business circles, this is "advertising" or "public relations." In the national security or information warfare communities, this is called "propaganda" or "psychological operations" or "disinformation."

Electromagnetic Pulse (EMP) or Directed-Energy Weapons (DEW): During the cold War, there was substantial civil defense preparation for potential Soviet attacks that consisted of atomic devices detonated at a certain altitude over the United States. The resulting blast of electromagnetic energy would literally and physically "fry" electronics, circuit boards, computer chips, and any other electronic item. Such weaponry exists today in terms of a relatively easy-to-conceal device that can be focus immense amounts of energy against a certain target or within a certain range, and from existing nuclear devices. *Given the many instances of suicide bombings associated with terrorism, using trucks (or even remote-controlled rockets and aircraft), picture a situation where a hostile organization was able to physically destroy a regional electronics infrastructure by using only a few precisely-placed atomic weapons. The result? Functional Paralysis through Nuclear Terrorism.*

Electronic Assassination: Imagine the surprise-you attempt to get a new drivers' license or open a bank account only to find out the computer has you listed as having been dead for five years! Or that you have been arrested for prostitution, drug-related crimes, and are badly in debt. Perhaps your employment record shows you having never graduated from high school, but yet you are working as a senior engineer for a major corporation? Have you ever lost a wallet and tried to prove your identity, get money, or make a purchase recently? *While "Hackers" and "The Net" were entertaining movies, they also presented some interesting and very real threat potentials. Anyone with the intent, a computer, modem, and some knowledge can do some very real damage to others. Again, we agree with the "old timers" who still keep receipts, invoices, and other important documents in the family safe-not that they don't trust computers, they don't trust the data on them or those using them. Today's large corporations and banks rely on computer databases for record-keeping. Modify that information, and you've changed the electronic life of a living, breathing person. Better yet—pay cash* *and avoid a commercial audit trail and subsequent invasion of your privacy by tracking your shopping habits.*

The threat is real and quite verifiable, just not seen as 'practical' for anything but the silver screen. It has many components-computers, modems, networks, hackers/crackers, cypherpunks, information security professionals, intelligence, info-terrorists, cyber-criminals, and so on.

The harbinger of this threat might be another country's intelligence establishment, terrorist group, organized criminal activity, or it may be the young professional across the hall in your apartment complex working for such an entity. "Knowledge Warriors" are those who have been reared during the Communications Revolution and are able to navigate the endless numbers of networks and databases with ease. Information Warriors, Knowledge Warriors, Info-Terrorists, Cyber-Warriors...the names are relatively the same. All are able to find whatever they seek. Some are able to carry out successful operations at the request of their employer. All seem to make good money doing it.

This new threat has not gone unnoticed by the United States Government. Key developments in recent history include the November, 1996, Defense Science Board report on *"Information Warfare: Defense"* that profiled the emerging threats and the pressing requirements to meet these threats to the defense infrastructure. This was confirmed by the October, 1997 release of *"Critical Foundations"* from the President's Commission on Critical Infrastructure (PCCIP), outlining to a greater degree the level of vulnerability of the United States and providing rudimentary guidance on countermeasures needed to remedy these issues. February, 1998 saw the National Infrastructure Protection Center (NIPC) stand up within the FBI as the national "watch center" for computer crime investigations, developing responses to information attacks, and providing intelligence analysis on the info/infra-structural warfare arena. In May, 1998, Presidential Decision Directive (PDD) 63 created the Critical Infrastructure Assurance Office (CIAO) to give oversight to the findings of the PCCIP report and work at the national level to foster the much-needed government-industry partnership in developing a defense of the infrastructures outlined in the PCCIP report of 1997. While a good start, it is too early to say whether or not such organizations will indeed be effective in the Information Warfare environment.

Unfortunately, the Congress has not had the foresight to further strengthen the American information infrastructures. In mid-1998, Congress eliminated the $68.8 million requested by the Department of Defense for information warfare defense and other activities. This was a complete about-face by Congressional offices when the Director of Central Intelligence and Director of the National Security Agency testified in both closed- and open-sessions on the dangers of information warfare and the critical requirement for appropriate funding levels. Instead, there was more Congressional attention paid to purchasing additional glitzy weapons platforms like the B-2, *which the Air Force has repeatedly said it does not need any more of,* to win the hearts and minds of their constituents. The classic Congressional-Pentagon debate over procurement being driven by constituents and not requirements is a historic one and while in no danger of disappearing, is frustrating those trying to effect an Information Defense with vaporware, paper exercises, and more bureaucracy.

The primary problem with strategic information warfare is that it is easy to do. As noted earlier, anyone with a computer or access to a network can search for, retrieve, or delete data on another network from anywhere in the world using any number of techniques, ranging from online tools such as search engines, jamming devices, viruses, Trojan Horses, or with traditional weapons like bombs. Or they could simply establish backchannel intelligence networks that are freely accessible to those knowing how to access them. And since the global information infrastructure has redefined the concepts of a "border" and national sovereignty, a Knowledge Warrior can easily cross into another nation electronically—without passport or warning—to accomplish what formerly had to be done in person at great risk. Further, since there are no front lines or "areas of (information) operations," anyone from anywhere can conduct information search, retrieval, or destroy operations.

February 1994 was our first observation of the ease of global information exchange and how it might benefit national security or intelligence communities around the world: From a dormitory room, college students were using a real-time "chat" service on the Internet known as IRC (Internet Relay Chat) to converse with a sixteen-year-old student in Sarejevo who watched "tanks rolling down my street and lots of soldiers walking and shouting outside." Those students on this particular "chat" service felt as though they were reading a modern techno-thriller instead of participating in an online discussion. Perhaps during future conflicts, this type of real-time information will prove vital in an unconventional information environment, conducted by those on-site and without military training or experience. With the glut of commercial "chat" rooms such as AOL and other mass-market online services, many discrete "backchannels" are made available to whomever needs to create them....and can be made just as secure (through the availability of unlimited channels to communicate in the clear) as an encrypted e-mail or phone communication.

This causes considerable headaches to the American intelligence establishments who are used to a Cold War-type of "early warning" of a hostile attack against the United States. For example, the North American Air Defense Command (NORAD) was tasked with using satellites, aircraft, and radar to instantly detect a Soviet missile launch. Back then, (and assuming the launch was detected quickly)the United States government had at least twenty to thirty minutes to decide how to respond from a land-based Soviet missile launched at the United States. Leading up to such an attack would be an escalating series of DEFense CONditions (DEFCON) that would determine the readiness posture of American forces worldwide Unfortunately, given the exponential number of methods, means, and locations to conduct an information attack, there are very few systems (classified or otherwise) that can provide this kind of advance warning of an organized Information Attack against the United States. In the same vein, the Pentagon (in conjunction with those non-Defense organizations developing the Information Defense of the country) is attempting to create a series of INFOrmation CONditions (INFOCON) that would be a graduated escalation of potential information warfare actions against the United States or its allies and interests and reminiscent of the nuclear DEFCON alerts. Time will tell if this INFOCON program will be limited to only Department of Defense components, or if this alerting system will be extended to the private sector that controls most of the world's information infrastructures.

Indeed, the Shakespearean paradox has come true, for today, "all the world's a stage" upon which the information warriors can act.

CHAPTER 5

Planning

One cannot refuse to eat just because there is a chance of being choked.

This stratagem suggests that one cannot forgo the benefits of a decision even if there is the slightest chance of failure. One has to eat to live even though there is a minute chance of choking. Today's organization has to experience the benefits of networking the employees and ultimately connecting to the Internet in order to conduct cyberbusiness and electronic commerce.

The basic tenets of information security worked well for the older and more stable mainframe technologies. The rules of Information Warfare engagement have changed in accordance with how technology has changed. Information security is still very much a management issue because the security program is dependent upon those they manage, namely people. First, upper management must be supportive of the security program. This aspect of business must filter from the top down. Secondly, the people who administer and program these systems must be knowledgeable relative to security issues and procedures. Their backgrounds must be checked to ensure they are not a risk to the data and the systems. After the basics are addressed (i.e., management and people) then technology issues can be addressed.

We continue with the analogy of the sword by making the comparison between the management and people aspects of security with the sword's handle of *tsuba*. It is the handle that is grasped by the hand that controls the sword's actions. Likewise the tsuba is the instrument that is used as the control over the security program. The hand that wields the sword is the hand that controls the actions of the people.

Sensei Sun

The TAO way of information warfare involves five measures. Measure first the risk, second the organization, third the telecommunications, fourth the platform, last the operating system. These measures must be analyzed at headquarters and are self-assessments.

The goal of the ruler should be to build an environment that is secure from all forms of attack. Just as the TAO of warfare is to win the fight by avoiding a fight, the goal of Information Warfare is not to symbolically "kill" the intruder but to fend off his every attempt to penetrate the system. Policies must be developed for all security matters that the ruler, chieftains, and warriors must follow.

The basic concept of assessing risk has not changed from the old technologies to the new. Assessing risk requires one to ask questions about the system and how it is protected. While the model of risk assessment has not changed, the questions one asks to assess risk have changed dramatically.

Yu Nix

Yes, as Sensei Sun says, first organization must assess its risk. Risk must be assessed by asking questions about the data; the organization; the operating system; the platform; and the telecommunications.

Sku Zi

In a risk assessment, one must question things such as: The physical security measures taken to protect the system. If there is physical security but there are no logical security measures such as strong passwords, there is risk.

If one uses passwords, but the passwords are weak, for example, not changed for long periods of time or protected from outside access, there is risk.

If one uses encryption, but the key length is short, there is risk.

Sensei Sun

A threat is like the wind; new threats can arise from any direction with intensity ranging from a gentle breeze to a great monsoon.

Yu Nix

There are many forms of measuring and assessing risk. Some are far too complex and lose their meaning. In a quantitative analysis of risk, if an item has a weighted risk factor of .02 while another is 1.2, what does it mean? And, who can translate such complex identifiers to those less-informed such as executives or end-users? One should simply define the risks of a system in a qualitative manner: high, medium, or low.

Com Pak

I agree. Further, a risk deemed to be of high measure should be given the most management attention, the most time, and the most budget in mitigating its potential effect.

Medium risks should be addressed next, and low risks last.

Sensei Sun

Organization—No other factor in this treatise is more important to the security posture of an organization than the organization itself. The commitment to security must begin and come from the top. Information security is not so much a technical issue as it is a management issue. Management must develop policies and procedures that are sound, concise, and strong enough to protect the people and systems.

If the ruler only speaks but does not practice what he speaks, the Chieftains will not have their hearts in the battle.

If the Chieftain only mimics the ruler and does not understand, the warriors will not have their hearts in the battle.

If the warrior's heart is not in the battle, the war will be lost.

The war must be led from the front, not the rear.

Com Pak

From Chieftain to warrior, all must be treated equally.

Lan Wan

Strategic assessments must be made within the organization. By the organization I mean all, the management, the systems personnel, and the users. Each group must have commitment and discipline for security.

Security is not the responsibility of one but of all.
The castle under siege must be fortified by all men, women, and children within its walls, if the battle is to be won.

Sku Zi

The organization must plan and be prepared for attack.

Com Pak

The lowest common denominator in information security is the risk associated with physical security.

Yu Nix

If the enemy can touch the computer, he can surely compromise the system.
Forget not to also apply physical security to the components in the field that support the system such as the distributed routers, hubs, and switches, and least of all do not neglect to consider the network wire itself.

Sensei Sun

The simplest and least sophisticated form of attack is denial of service. The simplest form of denial of service is to cut the wire. Cutting the network wire is like cutting a main artery. One denies service, the other denies life. In information security both are the same.

Com Pak

The cost of physical security should be proportional to the worth of the system and the information resident on the system.

Chieftains must know who has the keys to the gate. If the gatekeeper leaves the village, the keys must be retrieved from him and his name taken from the list of keyholders.

Sku Zi

Physical security also includes the environmental factors associated with using and prolonging the life of the equipment.

Com Pak

Equipment must not be used as a haven for food and drink. The equipment must be protected from the threat of fluctuations in electricity. A power source independent from the building's power source is needed for a graceful shutdown in the event of electrical loss.

Sensei Sun

After considering the physical security of the system, the integrity of the organization and the people who manage and operate the systems must be considered next. If they are not trustworthy, the system will be compromised in some manner. The least form of this type of compromise is neglect.

Management must be knowledgeable, trustworthy, decisive, and compassionate.

Yu Nix

Personnel that touch the data and the programs that control the data must be categorized and restricted according to their job responsibilities and their need to know.

First in the hierarchy are the system administrators. They know the internals of the system like a doctor knows the internals of the body. They must be trusted above all and their backgrounds should be scrutinized thoroughly.

Sku Zi

Next are the programmers. They should be permitted access to only that data for which they have a need to know.

All program development must be accomplished on test systems; production systems must never be used for development.

Lan Wan

In considering system testing, the test is severely hampered by using false data. In testing the goal should be to have as close to real data as possible.

Using false data in the test system will yield erroneous results in the production system.

Sensei Sun

At the core of the information security is the audit trail. Like a scout that tracks the enemy's movements, so too should security personnel follow the tracks in the audit logs and reports. Checking the system logs should be the first action of the day.

Com Pak

From Ruler to Chieftain,
From Chieftain to Warrior,
All must be committed to a common cause-the security effort.
The one with a weak door lock, the one with the weak password, the one with the weak management controls compromises the whole.

Sensei Sun

There are five qualities that can endanger the security posture of the organization:

<u>Recklessness with technology.</u> Caution must be exercised with the use of technology. One should not be quick to implement an X.0 software release, for serious problems can occur. One does not usually purchase a new model chariot and expect reliability. It must be tested and improvements made before it is worthy of use in battle.

<u>Ambivalence</u> on the part of Rulers, Chieftains and Warriors to the following sound security practices.

<u>Non-commitment by rulers.</u> The Ruler must have fought in the armor of a Chieftain and the Chieftain must have fought in the armor of a Warrior in order to truly understand the precepts of war.

<u>Cowardice to enforce rules.</u> If there is no punishment, there is no discipline.

– At the first instance of one's disregard in following the rules, he should lose his mouse-clicking finger.
– The second instance, he should lose his mouse hand.
– The third instance he should lose his Internet connection.

<u>Laziness</u> regarding attention to detail. If the Ruler permits the watchguard to sleep all will surely be lost.

Yu Nix

Rulers should not disembowel the Chieftain that delivers bad news, but rather the Ruler should disembowel the Chieftain that fails to deliver the bad news.

Sku Zi

One should confess early and often.

Com Pak

If the Ruler knows himself and his enemy, in battle he will never be in peril.
If the Ruler knows himself but not his enemy, in battle his chances are equal.
If the Ruler knows not of himself nor his enemy, in battle he surely will be in peril.

The Jewel of Knowledge

The jewel of the samurai lies in his knowledge. Today's technology requires continuous training for one to be effective. Training must be of high quality and continuous.

Sensei Sun

Training is the jewel of knowledge and is a power to be possessed by both the samurai and the information warrior.

Knowledge gained by training cannot be taken away from those who possess it.
Learning is a life-long experience.

Rulers should ensure that Chieftains are continuously trained and enriched in specific areas of expertise.

Chieftains should ensure warriors are crossed-trained on various duties because in the event one falls, another can take his place.

Just as data must be backed up, people must have back ups as well.

Sku Zi

Training must be of high quality and continuous.

If all people in the organization (rulers, Chieftains, and warriors) cease to learn, the organization is doomed.

Yu Nix

Unless warriors are able to use the technology, the skirmish will be lost.

Unless the Chieftain is able to understand and therefore, manage the technology, the battle will be lost.

Unless the ruler can envision new technologies, the war will be lost.

Sensei Sun

Next, the organization must affiliate itself with the Computer Emergency Response Team (CERT). The CERT organizations are familiar with the interworkings of the Internet. The CERT issues advisories regarding operating system and various other vulnerabilities. The CERT site has a listing and description of these vulnerabilities as well as program patches needed to fix these problems. The worth of the CERT organizations cannot be measured in gold. Familiarize yourself and your army with the CERT organization.

Telecommunications

Sensei Sun

The enemy is formless. He seeks your data and will embarrass your organization by stealing it. He will travel on your telecommunication lines and he will exploit your platform and operating system.

Information warfare can be waged from anywhere on the planet. No longer must a warrior travel to foreign soil to fight; the war is waged in cyberspace.

The organization should ensure that it gets the largest and fastest possible telecommunication lines because future applications will need this bandwidth.

Lan Wan

No longer is home soil sacrosanct.

Just as the quality of the roads were the key measure of the footsoldier's success, the telecommunication paths are the key to distributing the data, information, and knowledge-and therefore, the ultimate success of the organization.

Com Pak

The more complex the connectivity and number of interconnections, the better chance of success in distributing the data. However, the more intricate the network, the more fragile is the telecommunications infrastructure.
The more complex network will have more potential points-of-failure.
The more points-of-failure, the more potential points-of-entry for intruders.

Yu Nix

The wider the distribution, the greater the chance for unwanted interception by hostile forces.
Never send a password across an untrusted network in plain text. The password must be encrypted, unless, of course, a one-time password generator device has generated the password.

Sku Zi

Be wary of the network structure that depicts your lines leaving your site and entering a cloud. The cloud cannot be trusted. If you have hired an Internet Service Provider you must consider that unauthorized personnel can monitor your plain text transmissions.

Lan Wan

Ensure that all connection points in the network can operate at the fastest speed possible. The novice samurai should not purchase an ass since as soon as he learns to ride and becomes skillful in battle, he will rapidly outgrow the ass and realize that he needs a stallion. So it is with the speed of the network.

Sku Zi

The application's appetite for bandwidth is growing exponentially when compared to the bandwidth itself.

Sensei Sun

In setting up the network, first protect your perimeter with a firewall setup; next augment the firewalls with Intrusion Detection systems. Then protect the servers inside the perimeter.

He who believes that protection on the outside (the perimeter) is sufficient and no internal controls are necessary is a fool.

Protections on the inside are warranted as well. If the perimeter is strong and the servers within are weak, there is a great potential for internal compromise once the intruder has gotten past the perimeter controls.

If internal controls are weak, systems inside are vulnerable to attack from insiders.

This is the principle of the katana. The sword is flexible on the inside—*shingane*, and rigid on the outside—*hadagane*.

The Mirror of Introspection

The outdated mode of thinking with regard to Internet security was to have one firewall or choke point much like a castle's drawbridge. The most recent thinking is to have firewall devices augmented with Intrusion Detection systems positioned throughout the organization wherever they are needed to protect resources. Included in this discussion is the idea that the network or Intranet must be constantly monitored for intrusions since it is not a matter of if but when. The mirror of the samurai represents health. One needs only to look at oneself to assess one's health. This is true of the network or system that must be constantly monitored for assessment of its health.

Sensei Sun

One must heed the lesson of the mirror. The mirror is a reflection of reality.

Does not the wife assess her beauty daily in the mirror?

Does not the husband use the mirror for grooming?

The mirror is a reflection of truth and as such offers a model of introspection.

Yu Nix

Introspection means that one must constantly look at the status of firewalls as well as of internal network hosts and components.

One must constantly monitor firewalls to ensure that the intruder has not and cannot enter through this gate.

The internal host must constantly be monitored to ensure that they have not been configured in some manner that will compromise the firewalls.

Lan Wan

All warriors desire to be connected to the Internet. Ensure that their browsing is related to their duties and not a waste of time.

Sensei Sun

In order to protect one's network, one must think and act like the enemy, the hacker.
One must familiarize himself with the tools of the hacker and use these tools to assess his own network's security posture.

Sku Zi

The only true barrier in the world of electronics is an air gap.

Com Pak

Does the shark not taste its victim before it bites?
Does the enemy not send out spies before the war is waged?
Remember that the hacker will first probe your network and assess its weaknesses. You need to constantly monitor for these probes because an attack will surely follow. Precursors of attacks will usually come in the form of finger and ping attempts.

Platform

Platform relates to the hardware you are working with and trying to secure (i.e., mainframe, mini-computer, workstation, or personal computer). Each platform has its own unique set of security issues. The complexity associated with securing a specific platform is compounded by the operating system that is used with the platform.

Sensei Sun

The more platforms in one's overall configuration, the more complex the management, the more complex the security model.

Com Pak

Each platform, (i.e., mainframe, mini-computer, fileserver, and desktop) has its own unique security configuration and set of problems.

Sku Zi

To truly understand networking and telecommunications on the Internet, one must go back to the roots of the Internet. One must understand the Transmission Control Protocol/Internet Protocol (TCP/IP). One must also understand the fundamentals of the UNIX operating system.

Lan Wan

The UNIX operating system is where the CERT really illuminates the blackness of night. Familiarize yourself with the CERT advisories.

Sensei Sun

In UNIX, do not trust the "r" family of commands. The "r" means "remote" and permits both warrior and enemy alike to use the system from a remote location as though he was sitting in the same room with the equipment. The "r " commands are not to be trusted.

Lan Wan

Also, one should be wary of the services that use UDP (User Datagram Protocol). Since it is connectionless, it uses random ports and will compromise a packet filtering firewall scheme. TCP (Transmission Control Protocol) is connection-oriented and will use the trusted ports.

Sku Zi

Use TCP Wrappers on your UNIX host configurations so that only trusted warriors will have access to your systems.

Com Pak

Next, if one becomes familiar with the CERT advisories, he will notice that the UNIX sendmail service is very suspicious and should not be used if it is not necessary to do so.

Operating System

As stated above, the operating system provides the user with an interface and a means by which to control the hardware. Each operating system has its own strengths and weaknesses when it comes to security.

Sensei Sun

Ensure that the important software is the latest version as these usually have the latest security features included.

Yu Nix

Tests must continuously be conducted to ensure that software patches and versions do not negatively impact the intended security features of the software.

Lan Wan

Never trust the security of an operating system immediately after installing it. In the same way, do not trust the manufacturer's claims to product security but conduct your own independent verification...experiment with the software, talk to other administrators, and monitor the security information regarding that product from third-party assessors.

Sku Zi

Assess one's system against CERT and industry vendor bulletins to insure that the latest security patches are installed.

CHAPTER 6

Force

Hide your dagger behind a smile.

You can get close to your enemy by approaching with a smile while a dagger is hidden behind your back. This deception involves the hiding of one's true intentions. This ancient stratagem translates into modern day social engineering. Social engineering involves a number of methods by which the cyber-thief tries to obtain access to systems under the guise of a legitimate request. Over the telephone, the cyber-thief can request the root ID and password from an unsuspecting victim by posing as the computer center manager, for example. There are a number of phone scams where thieves will pose as Telephone Company technicians working on 800 number lines. They will request the victim to transfer the call to a '90' or '900' exchange, which ultimately connects them to an operator. They will then be able to make long distance calls at the victim's expense. Social engineering techniques are very effective and you need to train your people about social engineering techniques and to never give out information unless they know to whom they are giving it.

Tactical Considerations of Information Warfare

> *"Information-based Warfare is an approach to armed conflict focusing on the management and use of information in all its forms and at all levels to achieve a decisive military advantage especially in the joint and combined environment. Information-based Warfare is both offensive and defensive in nature-ranging from measures that prohibit the enemy from exploiting information to corresponding measures to assure the integrity, availability, and interoperability of friendly information assets."*
>
> **National Defense University**
> **(Excerpt of 1996 Working Definition)**

Information has always driven a war in the "traditional" sense of the word—campaigns waged by two or more entities on a battlefield with heavy weaponry, many casualties, and reliance on the wisdom of the commanding general and the information contained in his battle plans, maps, and remarks. War today is still the movement of forces to engage and overcome the enemy with the overwhelming application of firepower; however, today's military environment is driven more by information exchange at all levels of leadership than by tactics or the wisdom of a single general or his staff. Even the very essence of war-battles fought by soldiers with weapons-has evolved to a highly unconventional environment that knows no limits or boundaries; an environment where "force" or military-initiated efforts can be directed to a specific target (military or civilian) at a specific time and with a new series of weapons designed specifically for this new threat.

This reliance on battlefield communications is not without risk. Disruption of the battlefield C^3 infrastructure would most likely create sufficient havoc to allow an enemy the advantage and possibly achieve victory. Disruption or (preferably) denial of key information services is the primary goal of information warfare. Naturally, both sides will attempt this action, and the first to do so successfully gains the tactical advantage. In the opening rounds of Operation DESERT STORM, a successful 'communications attack' against Iraq resulted in both the enemy's inability to detect and coordinate a defense against the first wave of Allied strike aircraft and our subsequent first-strike successes on 16 January 1991. This was done with specially-modified weaponry that destroyed the internal circuitry of Iraqi C^3 systems, thus preventing enemy defenses from both detecting and responding to the coalition air attack and from warning Baghdad. Thus, Saddam Hussein was seriously hindered from coordinating a strong Iraqi defense (rumor has it that he did not know where his forces were, let alone those of the coalition!) by forcing him to use sub-optimal modes of communication. Even Sun Tzu promotes the use of deception to gain a battlefield advantage; his advice has withstood over a thousand years of review by military leaders.

During the American Civil War, the widespread reliance on the telegraph by the Union for tactical communications required sizable resources to protect its 15,000 miles of cabling from potential attack by the Rebel forces. By the end of the war, over half the Union Army was deployed in defensive positions around this vital line of communication. Had these telegraph cables-carrying some 3,300 coded messages daily-been severed, there would have been an unprecedented level of C^3-related confusion at all levels in the Union Army, and the outcome of the war might have been drastically altered. This was defensive information warfare waged over a hundred years ago, with inferior and less reliable technology than today's modern systems.

This radical change in the battle environment creates a major issue in establishing an all-encompassing definition for this new type of warfare. In the past, "warfare" connotated physical violence against two opposing combatants using military weaponry-gunpowder, tanks, or missiles-to win a battle through the effective use of force and maneuver. This is most evident in the writings of the leading scholar of the military arts, Prussian Karl von Clausewitz. His opus *On War* describes in painstaking detail how to wage a war of industry where such war material mentioned above were the primary tools in achieving a decisive victory. It is his doctrine that permeates and dominates many Western military schools of thought, including that of the United States. Unfortunately, in an information operation, the definition of a "war" expands to include a larger spectrum of conflicts, complete with many new combatants and the application of new weaponry in a non-physical military environment. Soldiers are now faced with the challenge of applying conventional information technology-computers and keyboards-in an unconventional military environment to counter traditional military technologies and tactics. As of this writing, each service has created (or is creating) specialized Info-War units to create doctrine for this new environment.

Former Secretary of the Air Force Sheila Windhall and then-USAF Chief of Staff General Ronald Fogelman saw information warfare as means to an end for both strategic and tactical purposes. While published from an Air Force perspective, the simple logic and examples are applicable to all military services and civilian organizations:

> Information warfare is any attack against an information function, regardless of the means. Bombing a telephone switching facility is information warfare. *So is destroying the switching facility's software. (Author's emphases)*
>
> Information warfare is any action to protect our information functions, regardless of the means. Hardening and defending the switching facility against air attack is information warfare. *So is using an anti-virus program to protect the facility's software. (Author's Emphasis)*

Conducting information operations using conventional military weaponry and doctrine worked well in the past, during situations where the available technology did not allow a "softer" method of attack, such as destroying an enemy's communications facility by air attack. A computer virus that causes an enemy's military C^3 systems to falter achieves the same result as a bombing, but without physical violence or destruction. Thus, a "soft" attack provides the same result but with a noticeable absence of violence, physical harm, or destruction. This subsequently brings war-the application of force-to a new level of effectiveness through the use of non-lethal technology, a by-product of information warfare. And, should such attacks fail, leaders can always resort to a "hard" attack using traditional tools of war.

We do not discount the usefulness of technology in warfare, but question the American soldier's reliance on it at the risk of not knowing his "backup" systems. For example, the military is growing reliant on GPS and computer-generated maps for the troops in the field. Paper maps are not as common as they used to be, as the information conveyed on them is only part of the total information presented on a GPS receiver or combat laptop. It is essential that the military continue to train its soldiers on "classic" military technology such as land navigation using maps, compass, and stars lest our combat units become paralyzed when its high-technology equipment fails.

In closing, ponder this. In a combat situation, a paper map with a bullet hole or some shrapnel through it is still a usable map and can be easily replaced if needed. A GPS receiver or combat laptop with a bullet hole or some shrapnel through it becomes an expensive, utterly useless trinket that is costly and difficult to replace. Bayonets (or swords, if we continue with the analogy) are not prone to regular breakdowns, frequent upgrades, new releases, power surges, EMPs, and a plethora of other high-technology vulnerabilities. And, in the wartime environment, a bayonet is more portable, easier to use, and has more personal applications for its user in than a laptop.

CHAPTER 7

The Other Side of Information Warfare

If you set your mind on your opponent's sword, your mind is consumed only with thoughts of his sword.

If you set your mind on timing, your mind is consumed only with thoughts of timing.

If you set your mind on your own sword, you mind is consumed with only thoughts of your sword.

In any case, your mind lingers, your actions falter, and you die.

Information is more than keyboards, mice, and modems. Unfortunately, it is not good business in today's world to market anything but firewalls, encryptors, and routers. Technology is a tool for good security, not a fire-and-forget solution. Education and training users and staff is imperative, as is an understanding that Information Warfare has other more sinister targets that can cause more damage than technological ones and are much more difficult—if not impossible—to defend against.

While most people consider "information warfare" to be waged exclusively against computers and infrastructures, this chapter is devoted to the most complex and vulnerable target of Information Warfare...the human brain. As mentioned earlier, anyone or anything can be a *victim* of information warfare, this chapter discusses the human being as a *target* of information warfare. Any organization that is run by or dependent on humans is inherently weak. From social engineering to criminal actions, humans—particularly Americans—are extremely trusting of someone being who he claims to be. Many consulting firms, the Big Five, and small companies offer security consulting services to install firewalls, encryptors, and guards for a company. The organization is then "blessed" or "certified" to be secure. Likewise, in the intelligence community, secret meetings are held in specially-designed facilities, with armed guards, two-person access rules, and all the protections in the world to insure the secrets stay secret. But secrets still get out.

Why?

All the best encryption, strongest firewalls, and all the security tools or money in the world cannot protect the human mind from making mistakes or errors in judgement. Similarly, there is no safeguard or firewall that can protect the most complicated—and vulnerable—information system in the world from a directed information attack. As such, it is a prime target for an adversary looking to achieve "information dominance" through actions directed against the human psyche instead of or in conjunction with attacks on physical information systems. If such actions can persuade, co-opt, trick, or influence, an enemy to unwittingly help them achieve their objectives through action (or inaction) the battle is half-won. Such actions are not limited to the Hollywood screen, but are very much a consideration in several nations around the world.

The remainder of this chapter is taken from an article written by Army Lieutenant Colonel Timothy L. Thomas, USA, (Ret.) in the Spring 1998 issue of *Parameters,* the quarterly professional journal of the US Army War College. The article is appropriately titled "The Mind Has No Firewall" and illustrates some very shocking and plausible attacks against the human mind for military applications. Note that while some of this article may seem far-fetched, it is included not only to discuss the "Other Side" of Information Warfare, but to demonstrate that our adversaries recognize and continue to research and explore this new realm of warfare.

> *"It is completely clear that the state which is first to create such weapons will achieve incomparable superiority."*
>
> **—Major I. Chernishev, Russian Army**

The human body, much like a computer, contains a myriad of data processors. They include, but are not limited to, the chemical-electrical activity of the brain, heart, and peripheral nervous system, the signals sent from the cortex region of the brain to other parts of our body, the tiny hair cells in the inner ear that process auditory signals, and the light-sensitive retina and cornea of the eye that process visual activity. We are on the threshold of an era in which these data processors of the human body may be manipulated or debilitated. Examples of unplanned attacks on the body's data-processing capability are well-documented. Strobe lights have been known to cause epileptic seizures. Not long ago in Japan, children watching television cartoons were subjected to pulsating lights that caused seizures in some and made others very sick.

Defending friendly and targeting adversary data-processing capabilities of the body appears to be an area of weakness in the US approach to information warfare theory, a theory oriented heavily toward systems data-processing and designed to attain information dominance on the battlefield. Or so it would appear from information in the open, unclassified press. This US shortcoming may be a serious one, since the capabilities to alter the data-processing systems of the body already exist. A recent news magazine highlighted several of these "wonder weapons" (acoustics, microwaves, lasers) and noted that scientists are "searching the electromagnetic and sonic spectrums for wavelengths that can affect human behavior." A recent Russian military article offered a slightly different slant to the problem, declaring that "humanity stands on the brink of a psychotronic war" with the mind and body as the focus. That article discussed Russian and international attempts to control the psycho-physical condition of man and his decision-making processes by the use of VHF-generators, "noiseless cassettes," and other technologies.

An entirely new arsenal of weapons, based on devices designed to introduce subliminal messages or to alter the body's psychological and data-processing capabilities, might be used to incapacitate individuals. These weapons aim to control or alter the psyche, or to attack the various sensory and data-processing systems of the human organism. In both cases, the goal is to confuse or destroy the signals that normally keep the body in equilibrium.

This chapter examines energy-based weapons, psychotronic weapons, and other developments designed to alter the ability of the human body to process stimuli. *One consequence of this assessment is that the way we commonly use the term "information warfare" falls short when the individual soldier, not his equipment, becomes the target of attack.*

Information Warfare Theory and the Data-Processing Element of Humans

In the United States the common conception of information warfare focuses primarily on the capabilities of hardware systems such as computers, satellites, and military equipment which process data in its various forms. According to Department of Defense Directive S-3600.1 of December 1996, information warfare is defined as "an information operation conducted during time of crisis or conflict to achieve or promote specific objectives over a specific adversary or adversaries." An information operation is defined in the same directive as "actions taken to affect adversary information and information systems while defending one's own information and information systems." These "information systems" lie at the heart of the modernization effort of the US armed forces and other countries, and manifest themselves as hardware, software, communications capabilities, and highly trained individuals. Recently, the US Army conducted a mock battle that tested these systems under simulated combat conditions.

US Army Field Manual 101-5-1, *Operational Terms and Graphics,* defines information warfare as:

> "actions taken to achieve information superiority by affecting a hostile's information, information based-processes, and information systems, while defending one's own information, information processes, and information systems." The same manual defines information operations as a "continuous military operation within the military information environment that enables, enhances, and protects friendly forces' ability to collect, process, and act on information to achieve an advantage across the full range of military operations. [Information operations include] interacting with the Global Information Environment . . . and exploiting or denying an adversary's information and decision capabilities."

This "systems" approach to the study of information warfare emphasizes the use of data, referred to as information, to penetrate an adversary's physical defenses that protect data (information) in order to obtain operational or strategic advantage. <u>It has tended to ignore the role of the</u>

human body as an information- or data-processor in this quest for dominance except in those cases where an individual's logic or rational thought may be upset via disinformation or deception. As a consequence little attention is directed toward protecting the mind and body with a firewall as we have done with hardware systems. Nor have any techniques for doing so been prescribed. Yet the body is capable not only of being deceived, manipulated, or misinformed but also shut down or destroyed—just as any other data-processing system. The "data" the body receives from external sources—such as electromagnetic, vortex, or acoustic energy waves—or creates through its own electrical or chemical stimuli can be manipulated or changed just as the data (information) in any hardware system can be altered.

The only body-related information warfare element considered by the United States is psychological operations (PSYOP). In Joint Publication 3-13.1, for example, PSYOP is listed as one of the elements of command and control warfare. The publication notes that "the ultimate target of [information warfare] is the information dependent process, whether human or automated . . . . Command and control warfare (C2W) is an application of information warfare in military operations. . . . C2W is the integrated use of PSYOP, military deception, operations security, electronic warfare and physical destruction."

One source defines information as a "non-accidental signal used as an input to a computer or communications system." The human body is a complex communication system constantly receiving non-accidental and accidental signal inputs, both external and internal. If the ultimate target of information warfare is the information-dependent process, "whether human or automated," then the definition in the joint publication implies that human data-processing of internal and external signals can clearly be considered an aspect of information warfare. Foreign researchers have noted the link between humans as data processors and the conduct of information warfare. While some study only the PSYOP link, others go beyond it. As an example of the former, one recent Russian article described offensive information warfare as designed to "use the Internet channels for the purpose of organizing PSYOP as well as for `early political warning' of threats to American interests." The author's assertion was based on the fact that "all mass media are used for PSYOP . . . [and] today this must include the Internet." The author asserted that the Pentagon wanted to use the Internet to "reinforce psychological influences" during special operations conducted outside of US borders to enlist sympathizers, who would accomplish many of the tasks previously entrusted to special units of the US armed forces.

Others, however, look beyond simple PSYOP ties to consider other aspects of the body's data-processing capability. One of the principal open source researchers on the relationship of information warfare to the body's data-processing capability is Russian Dr. Victor Solntsev of the Baumann Technical Institute in Moscow. Solntsev is a young, well-intentioned researcher striving to point out to the world the potential dangers of the computer operator interface. Supported by a network of institutes and academies, Solntsev has produced some interesting concepts. He insists that man must be viewed as an open system instead of simply as an organism or closed system. As an open system, man communicates with his environment through information flows and communications media. One's physical environment, whether through electromagnetic, gravitational, acoustic, or other effects, can cause a change in the psycho-physiological condition of an organism, in Solntsev's opinion. Change of this sort could directly affect the mental state and consciousness of a computer operator. This would not be electronic war or information warfare in the traditional sense, but rather in a nontraditional and non-US sense. It might encompass, for example, a computer modified to become a weapon by using its energy output to emit acoustics that debilitate the operator. It also might encompass, as indicated below, futuristic weapons aimed against man's "open system."

Solntsev also examined the problem of "information noise," which creates a dense shield between a person and external reality. This noise may manifest itself in the form of signals, messages, images, or other items of information. The main target of this noise would be the consciousness of a person or a group of people. Behavior modification could be one objective of information noise; another could be to upset an individual's mental capacity to such an extent as to prevent reaction to any stimulus. Solntsev concludes that all levels of a person's psyche (subconscious, conscious, and "superconscious") are potential targets for destabilization.

According to Solntsev, one computer virus that *could be* capable of affecting a person's psyche is Russian Virus 666. It manifests itself in every 25th frame of a visual display, where it produces a combination of colors that allegedly put computer operators into a trance. The subconscious perception of the new pattern eventually results in arrhythmia of the heart. Other Russian computer specialists, not just Solntsev, talk openly about this "25th frame effect" and its ability to subtly manage a computer user's perceptions. The purpose of this technique is to inject a thought into the viewer's subconscious. It may remind some of the subliminal advertising controversy in the United States in the late 1950s.

US Views on "Wonder Weapons": Altering the Data-Processing Ability of the Body

What technologies have been examined by the United States that possess the potential to disrupt the data-processing capabilities of the human organism? The 7 July 1997 issue of *U.S. News and World* Report described several of them designed, among other things, to vibrate the insides of humans, stun or nauseate them, put them to sleep, heat them up, or knock them down with a shock wave. The technologies include dazzling lasers that can force the pupils to close; acoustic or sonic frequencies that cause the hair cells in the inner ear to vibrate and cause motion sickness, vertigo, and nausea, or frequencies that resonate the internal organs causing pain and spasms; and shock waves with the potential to knock down humans or airplanes and which can be mixed with pepper spray or chemicals. These are called "non-lethal" weapons and technology.

With modification, these technological applications can have many uses. Acoustic weapons, for example, could be adapted for use as acoustic rifles or as acoustic fields that, once established, might protect facilities, assist in hostage rescues, control riots, or clear paths for convoys. These waves, which can penetrate buildings, offer a host of opportunities for military and law enforcement officials. Microwave weapons, by stimulating the peripheral nervous system, can heat up the body, induce epileptic-like seizures, or cause cardiac arrest. Low-frequency radiation affects the electrical activity of the brain and can cause flu-like symptoms and nausea. Other projects sought to induce or prevent sleep, or to affect the signal from the motor cortex portion of the brain, overriding voluntary muscle movements. The latter are referred to as pulse wave weapons, and the Russian government has reportedly bought over 100,000 copies of the "Black Widow" version of them.

However, this view of "wonder weapons" was contested by someone who should understand them. Brigadier General Larry Dodgen, Deputy Assistant to the Secretary of Defense for Policy and Missions, wrote a letter to the editor about the "numerous inaccuracies" in the *U.S. News and World Report* article that "misrepresent the Department of Defense's views." Dodgen's primary complaint seemed to have been that the magazine misrepresented the use of these technologies and their value to the armed forces. He also underscored the US intent to work within the scope of any international treaty concerning their application, as well as plans to abandon (or at least redesign) any weapon for which countermeasures are known. One is left with the feeling, however, that research in this area is intense, if not government-sponsored, than certainly privately-funded or academic. A concern not mentioned by Dodgen is that other countries or non-state actors may not be bound by the same constraints. It is hard to imagine someone with a greater desire than terrorists to get their hands on these technologies. "Psycho-terrorism" could be the next buzzword.

Russian Views on "Psychotronic War"

The term "psycho-terrorism" was coined by Russian writer N. Anisimov of the Moscow Anti-Psychotronic Center. According to Anisimov, psychotronic weapons are those that act to "take away a part of the information which is stored in a man's brain. It is sent to a computer, which reworks it to the level needed for those who need to control the man, and the modified information is then reinserted into the brain." These weapons are used against the mind to induce hallucinations, sickness, mutations in human cells, "zombification," or even death. Included in the arsenal are VHF generators, X-rays, ultrasound, and radio waves. Russian army Major I. Chernishev, writing in the military journal *Orienteer* in February 1997, asserted that "psy" weapons are under development all over the globe. Specific types of weapons noted by Chernishev (not all of which have prototypes) were:

A psychotronic generator, which produces a powerful electromagnetic emanation capable of being sent through telephone lines, TV, radio networks, supply pipes, and incandescent lamps.

An autonomous generator, a device that operates in the 10-150 Hertz band, which at the 10-20 Hertz band forms an infrasonic oscillation that is destructive to all living creatures.

A nervous system generator, designed to paralyze the central nervous systems of insects, which could have the same applicability to humans.

Ultrasound emanations, which one institute claims to have developed. Devices using ultrasound emanations are supposedly capable of carrying out bloodless internal operations without leaving a mark on the skin. They can also, according to Chernishev, be used to kill.

Noiseless cassettes. Chernishev claims that the Japanese have developed the ability to place infra-low frequency voice patterns over music, patterns that are detected by the subconscious. Russians claim to be using similar "bombardments" with computer programming to treat alcoholism or smoking.

The 25th-frame effect, alluded to above, a technique wherein each 25th frame of a movie reel or film footage contains a message that is picked up by the subconscious. This technique, if it works, could possibly be used to curb smoking and alcoholism, but it has wider, more sinister applications if used on a TV audience or a computer operator.

Psychotropics, defined as medical preparations used to induce a trance, euphoria, or depression. Referred to as "slow-acting mines," they could be slipped into the food of a politician or into the water supply of an entire city. Symptoms include headaches, noises, voices or commands in the brain, dizziness, pain in the abdominal cavities, cardiac arrhythmia, or even the destruction of the cardiovascular system.

There is confirmation from US researchers that this type of study is going on. Dr. Janet Morris, co-author of *The Warrior's Edge,* reportedly went to the Moscow Institute of Psychocorrelations in 1991. There she was shown a technique pioneered by the Russian Department of Psycho-Correction at Moscow Medical Academy in which researchers electronically analyze the human mind in order to influence it. They input subliminal command messages, using key words transmitted in "white noise" or music. Using an infra-sound, very low frequency transmission, the acoustic psycho-correction message is transmitted via bone conduction.

In summary, Chernishev noted that some of the militarily significant aspects of the "psy" weaponry deserve closer research, including the following nontraditional methods for disrupting the psyche of an individual:

ESP research: determining the properties and condition of objects without ever making contact with them and "reading" peoples' thoughts

Clairvoyance research: observing objects that are located just beyond the world of the visible—used for intelligence purposes

Telepathy research: transmitting thoughts over a distance—used for covert operations

Telekinesis research: actions involving the manipulation of physical objects using thought power, causing them to move or break apart—used against command and control systems, or to disrupt the functioning of weapons of mass destruction

Psychokinesis research: interfering with the thoughts of individuals, on either the strategic or tactical level

While many US scientists undoubtedly question this research, it receives strong support in Moscow. The point to underscore is that individuals in Russia (and other countries as well) believe these means can be used to attack or steal from the data-processing unit of the human body.

Solntsev's research, mentioned above, differs slightly from that of Chernishev. For example, Solntsev is more interested in hardware capabilities, specifically the study of the information-energy source associated with the computer-operator interface. He stresses that if these energy sources can be captured and integrated into the modern computer, the result will be a network worth more than "a simple sum of its components." Other researchers are studying high-frequency generators (those designed to stun the psyche with high frequency waves such as electromagnetic, acoustic, and gravitational); the manipulation or reconstruction of someone's thinking through planned measures such as reflexive control processes; the use of psychotronics, parapsychology, bioenergy, bio fields, and psychoenergy; and unspecified "special operations" or anti-ESP training.

The last item is of particular interest. According to a Russian TV broadcast, the strategic rocket forces have begun anti-ESP training to ensure that no outside force can take over command and control functions of the force. That is, they are trying to construct a firewall around the heads of the operators.

Conclusions

At the end of July 1997, planners for Joint Warrior Interoperability Demonstration '97 "focused on technologies that enhance real-time collaborative planning in a multinational task force of the type used in Bosnia and in Operation Desert Storm. The JWID '97 network, called the Coalition Wide-Area Network (CWAN), is the first military network that allows allied nations to participate as full and equal partners." The demonstration in effect was a trade fair for private companies to demonstrate their goods; defense ministries got to decide where and how to spend their money wiser, in many cases without incurring the cost of prototypes. It is a good example of doing business better with less. Technologies demonstrated included:

Soldiers using laptop computers to drag cross-hairs over maps to call in airstrikes

Soldiers carrying beepers and mobile phones rather than guns

Generals tracking movements of every unit, counting the precise number of shells fired around the globe, and inspecting real-time damage inflicted on an enemy, all with multicolored graphics

Every account of this exercise emphasized the ability of systems to process data and provide information feedback via the power invested in their microprocessors. The ability to affect or defend the data-processing capability of the human operators of these systems was never mentioned during the exercise; it has received only slight attention during countless exercises over the past several years. The time has come to ask why we appear to be ignoring the operators of our systems. Clearly the information operator, exposed before a vast array of potentially immobilizing weapons, is the weak spot in any nation's military assets. There are few international agreements protecting the individual soldier, and these rely on the good will of the combatants. Some nations, and terrorists of every stripe, don't care about such agreements.

This article has used the term data-processing to demonstrate its importance to ascertaining what so-called information warfare and information operations are all about. Data-processing is the action this nation and others need to protect. Information is nothing more than the output of this activity. As a result, the emphasis on information-related warfare terminology ("information dominance," "information carousel") that has proliferated for a decade does not seem to fit the situation before us. In some cases the battle to affect or protect data-processing elements pits one mechanical system against another. In other cases, mechanical systems may be confronted by the human organism, or vice versa, since humans can usually shut down any mechanical system with the flip of a switch. In reality, the game is about protecting or affecting signals, waves, and impulses that can influence the data-processing elements of systems, computers, or people. We are potentially the biggest victims of information warfare, because we have neglected to protect ourselves.

Our obsession with a "system of systems," "information dominance," and other such terminology is most likely a leading cause of our neglect of the human factor in our theories of information warfare. It is time to change our terminology and our conceptual paradigm. Our terminology is confusing us and sending us in directions that deal primarily with the hardware, software, and communications components of the data-processing spectrum. We need to spend more time researching how to protect the humans in our data management structures. Nothing in those structures can be sustained if our operators have been debilitated by potential adversaries or terrorists who—right now—may be designing the means to disrupt the human component of our carefully constructed notion of a system of systems.

CHAPTER 8

Armed Struggle

Kill with a borrowed knife.

This stratagem is the ultimate deception. In ancient times, one would kill with a borrowed knife so that the owner would be blamed and not the true murderer. There are many examples of this type of deceptive behavior on the Internet. One of the primary examples is the E-mail message. It is very easy to spoof the "From" address of e-mail so that the message appears as if it has come from just about anybody. If you are not using digital signature authentication, you cannot assume that the message is from whom it says it is from.

Special Topics of Information Warfare

The following Special Topics of Information Warfare discuss specific areas of concern or the armament of battle. Topics included are e-mail, computer viruses, wireless communications, programmers, projects, chieftains and contractors. These topics are based on experience-blood, sweat, and fears.

E-mail

No other application has offered more in terms of productivity and utility, while at the same time causing inordinate concern in the security arena. Almost everyone has, is, or will be using e-mail of one sort or another. No one will argue about the utility of e-mail but the concern over e-mail privacy and security is rampant. For one, e-mail is easily "spoofed" that is to say that one can disguise oneself as just about anyone and send off an e-mail message. This vulnerability causes many celebrities and public figures to receive vulgar e-mail messages or even worse, death threats. Law enforcement officials then have an almost impossible task to find the culprits. Secondly, e-mail messages can be used as the weapon of choice for denial of service attacks. The authors have downloaded "hacker" software capable of attacking a mail server by sending 10,000 e-mail messages from an anonymous source with the click of a mouse. Lastly, e-mail messages are being used as delivery vectors for viruses. This attack is accomplished by sending the "infected file" (virus) as an attachment to an e-mail message. The recipient launches the attachment in a viewer, and the virus is activated.

Sensei Sun

Everyone has used an E-mail package. Therefore, just as everyone has a rectum, everyone has an opinion as to which E-mail package is best. No one will be happy with the corporate decision to standardize on one e-mail package. Absolutely no one e-mail product will have all the features or security that the users need and want.

Lan Wan

E-mail will acquire a life of its own and truly requires a lesson in and of itself. Some of the e-mail lessons and rules include the following:

Make sure you are sending the e-mail message to the right person. There may be five Wongs in your organization. It would be ill advised to send a personal message concerning one Wong to the wrong Wong. And two Wongs don't make a right.

E-mail should be used for official company business. The ruler should not be on the receiving end of an errant message regarding the kettleware party you are having tonight.

E-mail messages should not be archived for long periods of time. The messages should only be maintained only long enough for recovery time (2 weeks is a good rule of thumb). If kept longer, the lawyers will want your files for litigation purposes. What you do not have in your possession and is unrecoverable will not get you in trouble.

E-mail can be easily forged. A message, which appears to have been sent from the ruler, might, in fact, have been sent from the court jester.

Set limits on the size of attachments one can send via E-mail. Multi-megabyte file transmissions can choke a network faster than the butcher can ring a pheasant's neck, especially if the E-mail message is sent to multiple parties.

Ensure E-mail administrators are trustworthy since they can take action to read the e-mail of all. If the message is important the delivery hierarchy should be:

First—in person
Second—over the telephone.
Third—via Federal Express.
Fourth—via fax.
Fifth—via e-mail.

If one is not willing to sign his E-mail message with his own name he is not trustworthy.
Anonymous remailers are used by vermin and snakes. He that refuses to sign his name and uses an anonymous remailer is lower than whale dung.

Computer Viruses

The then National Computer Security Association (NCSA) issued the results of a computer virus study in 1996. The data is astonishing. NCSA found that, "Virtually all North American corporations and other large organizations (98%) have experienced computer virus infections first hand. As of early 1996, about 90% of all organizations with more than 500 PCs experience a computer virus encounter or incident each month. The chance of experiencing a computer virus encounter or incident in early 1996 appears to be about one chance per 100 PCs per month."

The NCSA report concluded that the bottom line is that things are getting worse. Over 90% of the respondents considered that computer virus problems in the computing industry are about the same or worse compared to this time last year, whereas almost no one (9.7%) thought that the problem has gotten any better:

The Center for Disease Control has a classification system for identifying the severity of the various viral and bacterial strains. The least innocuous biological viruses are designated as Level 1 types of bugs and the most deadly and likewise, most interesting are the Level 4 types of viruses (e.g., Ebola). In order to handle Level 4 viruses, the laboratory workers must wear protective space suits and breathe oxygen provided through thin, straw-like tubing piped in from the outside. They protect their extremities with double layers of gloves and boots. There is even an international symbol for toxic materials—the trefoil—to alert scientists worldwide that extremely hazardous materials are present. At the CDC, Level 4 virii have their own trefoil label which identifies that they are biohazardous material. Some Level 4 viruses—like Ebola—have a fatality rate in the 90-plus percentile.

Computer viruses can be categorized in the same manner. Sensei Sun has also provided a four-tiered taxonomy for computer viruses. **Level 1** (innocuous), **Level 2** (message bearers), **Level 3** (modifiers) and **Level 4** (deadly). Level 4 computer viruses can be considered as i/o-hazardous materials.

Sensei Sun

Just as a virus can make a human ill, the computer virus can make the machine ill.

The taxonomy of i/o hazards is:

Level 1—viruses that do not cause any damage are considered innocuous,

Level 2—viruses that contain a humorous message but cause no damage,

Level 3—viruses that are more dangerous than the first two levels and cause unauthorized changes to data, and

Level 4—viruses that cause a system catastrophe usually without warning.

Com Pak

The computer virus, like its biological brethren, is called a virus because of its propensity to replicate itself and cause harm.

Yu Nix

Yes, and one must remember that the computer virus also needs a vector with which to transmit itself from host to victim. The vector can be a floppy diskette, a program copied over a network, or a small macro program embedded in a document.

Lan Wan

Some viruses are benign and cause mischief; some are malicious and cause catastrophic failure.

Com Pak

Each virus has a place where it hides in the computer; this is called the virus' signature. The virus can hide in memory, can infect the boot sector or be mailed as macrocode appended to an e-mail message.

Sensei Sun

The chieftain is truly wise if he uses two different virus protection programs at his site. The success of a virus program is in its ability to add new virus signature databases at routine intervals of time (i.e., at least quarterly). If one uses two different virus protection programs,

chances are that while one is outdated, the other will be current with regard to the latest virus signatures.

Yu Nix

Check everything you put into your computer including shrink-wrapped software, and especially check everything downloaded from the Internet.

Com Pak

Some computer platforms are more apt to generate computer viruses.

Sensei Sun

Viral code is rampant on operating systems that:
allow the core operating system to run in an unprotected domain
allow system files and data to be violated
allow unrestricted access to raw disk devices
allow unauthorized access to programs or files
One should always practice safe HEX.

Wireless Communications

Wireless communications (such as cellular telephones and cordless telephones) are among some of the easiest of eavesdropping targets. Given the right conditions, motivation, and resources, reception of these conversations can be very clear. Cellular telephone communications can be intercepted over hundreds of square miles while home cordless telephones can be intercepted up to one mile away. Other telephones, which are susceptible to monitoring, include commercial airline and rail telephones, ocean liner phone calls, long distance calls sent via satellite and long distance calls sent via microwave radio links. Even the lowly paging device is not immune to the threat of interception.

In addition to being an espionage windfall, monitoring of these transmissions has also become a national pastime with opportunistic hobbyists. The interception and disclosure of wire, oral, or electronic communications is illegal under 18 USC 2511 and can result in civil or criminal sanctions. There is no deterministic method to identify that one's transmissions are being intercepted over the airwaves. Without some form of disclosure, enforcement and prosecution of casual interceptions is nearly impossible. New techniques in cellular eavesdropping include computer assisted, totally automated monitoring which allows monitoring of specific phones, 24-hours a day, from cell to cell, without human assistance.

The following countermeasures identify prudent security practice for existing analog (regular) cellular and cordless phones, good security (marginal cost), better security (more expensive,), and best security (most expensive) measures that can be taken with regard to wireless communications.

Sensei Sun

If using a cordless phone, you might as well shout your battle strategy into your adversary's camp,

If using a cell phone, you might as well shout your battle strategy in the Town Square of your adversary's capital city.

Security Tips for Analog (Regular Cellular and Cordless Phone)

Protect your car telephone conversations by arranging to call on a number, which is not answered with a company/office name or other identifying information.

Use first names and code words to identify special projects.

Speak in general and uninteresting terms.

Switch to your regular "wired" telephone for increased-though not absolute-security.

When traveling, don't pull over in order to conduct a conversation. You can achieve some minor security protection by virtue of the fact that cellular transmissions will be transferred from "cell to cell" (approximately every few miles) if the person is traveling while conversing.

Ensure that the other people involved in the conversation are taking similar security precautions. For example, if you are speaking from a regular "wired" telephone, and one of the other parties is speaking from a cellular phone, the conversation can be monitored.

If you must have a cordless phone, buy one which operates in the 900 MHz frequency range or higher using digital spread spectrum technology.

Good Security

The "hobbyists" monitor the airwaves for regular (analog) cell phone and cordless transmissions. Some of the new digital cellular phones provide better security because they must convert analog signals into digital signals prior to transmission. Scanners capable of intercepting and interpreting digital transmissions (converting the signals from digital to analog) are expensive and therefore are not popular with the general public. The consensus among security professionals is that digital transmissions will remain safe for some years to come since the majority of devices will continue to be analog and therefore easier and cheaper to intercept.

Digital cellular phones are considered "passive" devices since the user does not have to actively place the phone in the digital transmission mode. Digital phones, however, can fade from digital to analog and back again depending on the capabilities of the "cell" one is traveling through. "Cells" capable of handling digital transmissions are not a problem in the metropolitan areas but may be a problem in outlying areas. Also, both ends of the conversation must be protected in a similar manner.

Better Security

For additional cellular protection, consider having your telephone's security enhanced via the use of new "scrambling" devices. Some companies offer a modification to existing cellular telephones whereby special enhancements such as external slip-on sleeves or internal chips that scramble transmissions are available at moderate cost.

There are some drawbacks to the use of the "scrambler" technology, such as both ends of the cellular conversation must be protected (scrambled), not all areas of the country can accommodate this technology as yet, and the device is not passive and requires that the user dial a specific sequence of keystrokes in order to toggle between "scrambled" and unscrambled transmission modes.

Best Security

If you have need for unequivocal security, only encryption will provide the best end-to-end security but also has the highest cost. Military-grade telephone encryption is not available to the general public and is used only in special cases. One of the major communications companies advertises cellular phone products based on the controversial Clipper chip encryption technology. Another of the major communications companies has announced its encrypted cellular phone product line in early February 1997.

Programmers

Sensei Sun

Like warriors, the programming staff will exhibit battle fatigue as follows:

> ***If the programmer is slumped over his keyboard, he or she is either extremely tired, coded-out, or dead.***
>
> ***If the programmer's reading lamp is on all night, he is behind schedule.***
>
> ***If the programmer's reading lamp is off all night, he is behind schedule.***

When the programmer has begun to drink the oil from the lamp, he is back on schedule.

Projects

☯ Lan Wan

If the project is on schedule, it is behind schedule.

User requirements are never properly defined.

If one cannot test a system with live data, he should test with data as close to live as possible. For example, a financial system that is tested with bogus values in the test environment will most likely yield bogus results in the production environment.

Getting users to test the system is like pushing on a rope, squeezing a nickel from a miser, or teaching a dog to play the lyre.

There are six phases to the project:

At first the tribe is enthusiastic,

Second, the tribe becomes disillusioned, [usually when they find out that they have to do the project in an abbreviated or impossible amount of time].

Third, panic sets in,

Fourth, a designated Chieftain searches for the guilty among the tribe,

Fifth, there is punishment of the innocent,

Lastly, Praise and Honors are heaped upon the non-participants.

Chieftains

☯ ***Yu Nix***

If the chieftain brags about his accomplishments, he is afraid of losing his job.

If the chieftain brags about the accomplishments of his warriors, he is truly wise and worthy of warrior loyalty.

For the warrior to follow, the chieftain's actions must speak louder than words.

Show me a leader that is willing to take a risk and I'll show you a leader.

Sensei Sun

Leadership must:

Lead people to insights about themselves, their dreams, and their roles.

Lead people from dependence to independence and then to interdependence.

Empower, inspire, and release people to fulfill their purpose and destiny.

Be earned not dictated.

Motivate people to be interdependent, committed and secure in transparent relationships.

Not rule through fear but through cooperation and support.

Not create a climate where leaders at all levels don't feel threatened by new ideas and new innovations.

Provide rewards for those who foster good ideas.

Be good stewards of its nation's treasure—its young men and women.

Practice self-discipline.

Never tolerate any breach of integrity.

Maintain accountability.

Maintain trust.

Maintain zero tolerance for any harassment or prejudice based on race, religion, ethnic origin, or sex.

Place service above self.

Contractors

Sensei Sun

Dealing with outside contractors is like conducting business with village prostitutes. They only want your money and when the business has been transacted, you realize you would have been better off taking matters into your own hand.

Com Pak

Hold your contractors to the same standards of security assigned to those employees they are working with from your own company. They must sign the same security forms, receive appropriate security briefings, and comply with nondisclosure agreements.

CHAPTER 9

Call to Arms

A fall into a ditch makes you wiser.
Learn, Learn, Learn. Train, Train, Train.

"The More You Sweat in Training,
The Less You Bleed in Combat"

- US Navy SEALS

Enough Said.

"War, in common with sport, has the characteristic that what worked well yesterday may not work well tomorrow, precisely because it worked yesterday."

Edward N. Luttwak Strategy:
The Logic of War and Peace (1987)

In the last 15 years America has been invaded by what has been known as information technology. Like the body snatchers of *Alien* that penetrated deep into the human body, computers and communication technologies have penetrated deep into our lives. Unfortunately, the *Alien* metaphor may not be apt since for the most part we have willingly *invited* this force into our homes and not had such forces directly forced upon us. We invited these technologies into our homes and our businesses because they allowed us to do things faster, to do things better and to do things cheaper. Among other things these technologies have reduced the cost of running a home, made our businesses more competitive, opened new markets by bringing buyers and sellers closer together, and expanded the horizons of our students not to mention adding entertainment value to our lives. This remainder of this chapter is taken from remarks to the United States Senate on 12 October 1998 by Senator Robert Kerrey (D-NE).

The good news of computer and associated communication technology have been offset by our growing dependence. To see how much we are dependent one need only look at the high level of concern surrounding the Y2K problem. Computer software is written so that at a second after midnight on January 1, 2000, while hundreds of millions of humans will be celebrating the end of an old millennium and the beginning of a new, our computers will act as if it is January 1, 1900. To the machines this will be the equivalent of a "daylight saving century." To some this is the beginning of a humorous and good news story: No income tax, a chance to correct the terrible mistakes of the past 100 years, and so forth. However, for those who operate our banking, emergency response, air traffic control, and power systems this will be nothing to laugh at. So dire are the predictions of some who understand how dependent on computers and software we have become that they talk as though they are storing up food and medical supplies just in case.

None of this would have happened if the century had ended 20 years earlier because computers, chips, and microprocessors were not yet running things. Twenty years ago we heard how computers were going to change the world. In 1983, portable computers were available only to those with strong backs or a forklift to carry such large, cumbersome (yet portable!) systems. E-mail was in its infancy. The Internet was 10 years away from its grand opening to the public. Software was built into mainframes and was available to those who knew how to navigate the procession of prompts and confusing signs. Speed was a snails' pace. Capacity was like a rain drop in the desert.

What happened in the past 20 years is that we were thirsty for the things a computer could do for us. Rapid and accurate calculations enabled even small businesses to get costs under control. Personal computers empowered us. Desktops enabled us. Laptops liberated us. Decision making—once driven from the top down by men and women with MBA degrees–has been distributed outward and downward to those with the technical know-how to not only navigate the complex systems of today, but to see the interaction these systems made to the global marketplace and "system of systems."

Today, any PC or Macintosh with average speed and power with state-of-the-art connectivity makes its user a publisher, broadcaster, editor, opinion maker, and analyst of large amounts of previously confusing data. Such data can be made instantly available to anyone, anywhere, at any time.

Advances in computer and telecommunications technology have spurred change and growth in our economy. These changes have generated wealth and jobs by creating new businesses and destroying old ones. Market-oriented businesses have had to adjust or perish. Public institutions, because of the nature of democracy–in other words, *majority rules* but *narrow interests win* elections–have been changing much more slowly. Slowly but surely the work of transferring knowledge from a teacher to a student is being done with the assistance of computers, software, and new systems where new skills are needed.

In fact, nowhere are the changes of the computer age more pronounced than in our military and intelligence organizations. Computers and communication technologies have made America's fighting forces stronger and more effective. We should be proud of the men and women who have trained and prepared themselves to take advantage of these new tools.

However, we also need to be alert to a hard truth: With strength comes vulnerabilities. Just as Achilles was held by his heel as he was dipped in the potion that made him unbeatable, we need to be alert to those small spaces where a determined enemy could do us great harm. If we are to maintain our economic success and provide the security our citizens expect and deserve, we must as a nation turn to address our weaknesses.

The ability of people to use information technology to reach into our homes and to amass vast amounts of personal data threatens our sense of privacy. The omnipresence of this technology has caused our society to develop a dependency on silicon chips and the wires that connect them. And, the connectivity that now brings us so many benefits may also be a vulnerability that nations and terrorists could use to threaten our security.

We have been blessed by our dominance in high-technology industries and in our society's acceptance of new information technology. Information systems are the backbone of America's telecommunications and electrical power grids, banking and finance systems, our transportation systems, broadcast and cable industries, and many other businesses besides. They have helped American workers become more productive, have brought new efficiencies in the use and distribution of resources, and have helped our nation grow to be the most advanced and competitive economy in the world.

We owe a large part of that success to the ingenuity, perseverance, and vision of America's information technology companies and their employees. The story of how computer companies started in garages can grow into multibillion dollar corporations is almost legendary. An industry virtually non-existent twenty five years ago has brought enormous wealth and opportunity to thousands of Americans.

Information technology has transformed our nation's economy, and, as we enter into the 21st century, our nation's livelihood will depend on continued development of this industry. But the wonder of this technology is how its success has brought extraordinary changes to other aspects of our lives.

Modern information technologies provide us with unheard-of opportunities in education, business, health care, and other life-enriching areas. Information technology empowers people to continue their education and upgrade their skills throughout life. Education no longer ends at the schoolhouse door. In addition, new technologies are extending lifesaving medical care to remote rural areas and promoting healthy communities across the country. These new avenues to information better inform our electorate, and the improved means of communication make it far easier for individual citizens to express their views to the general public and to their elected representatives.

In combination, these technological benefits allow people—both young and old—to develop new skills, explore new interests, and improve their lives. America's technological strength is the envy of nations around the globe. But that strength, if not understood and protected, may also be our Achilles' heel.

We have been blessed this year with a number of warnings about this grave and far-reaching threat. We have been blessed with warnings about the interdependence of our information infrastructures, the interlocking network that can make local hospitals and airports victims just as easily as multinational corporations and media conglomerates.

We need to heed the warning and respond to this danger. In late 1998, the media reported that the electronic mail programs the vast majority of Americans use had vital, hidden flaws. Simply opening an e-mail message could unleash a malicious virus and allow that virus to freeze a computer, steal data, or erase a hard drive. While there are some people in the United States who still do not use e-mail, it is a fact that our society today relies upon electronic mail for use in government and commercial communications, for business management and project coordination, and personal entertainment and missives. A malicious person could potentially have used these flaws to blackmail people or companies, to disrupt government and commercial activity, or to sabotage civilian or military databases.

During 1998, the Galaxy-4 satellite orbiting more than 22,000 miles above the state of Kansas began tumbling out of control. It was the worst outage in the history of satellites. By conservative estimates, more than 35 million people lost the use of their pagers, including everyone from school children and repairmen to doctors, nurses, and other emergency personnel. Other systems were affected as well, including gas station pay-at-the-pump systems for a majority of the country.

All of that was the result of one small computer on one small satellite 22,000 miles in outer space.

Early 1998 had the United States again in the middle of a very tense standoff with Saddam Hussein. And we were able to track an attack on the Pentagon's computer system to a site in the Middle East, in the United Arab Emirates. There was a legitimate question at the time: *Was this an act of war? Was it a terrorist?* Or was it, as it turned out to be, *teenage joyriders* inappropriately and illegally using their home computers? The implications of an effective attack against our military's information systems would be devastating during a time of crisis. This attack failed, but will we be as fortunate in the future?

We do not think these incidents are a statement only about software companies, the satellite industry, or teenage computer aficionados. These incidents are a warning–loud, clear, and wide–about the dependence of the American economy and the American people on information technology. Our use of information technology has helped us achieve and maintain our status as the world's strongest nation. But our dependence on information technology also brings exploitable weaknesses that, like the Lilliputians to the giant Gulliver, may enable our weaker adversaries to cause great damage to our nation.

In Jonathan Swift's tale, the Lilliputians used their mastery of mathematics and technology to defeat their much more powerful adversary Gulliver. Today, weaker adversaries may use their mastery of information technology to invade our privacy, steal from our companies, and threaten our security. Indeed, there are even fourteen-year-old teenager "hackers" who are teaching advanced UNIX programming and making a tidy profit doing so.

The revolution in information technology has propelled the United States to an unparalleled position in the global economy. The principles of freedom and democracy that we champion are ascendant throughout the world.

We have the world's largest economy, and we trade more than any other nation. Our military strength, in conventional and nuclear terms, is greater than that of any other nation. In short, we are the sole remaining superpower in the world.

And yet, we still find ourselves vulnerable to individuals or groups—terrorists, criminals, saboteurs—who have a fraction of the manpower, weaponry, or resources we possess. In many ways, we are a technological

Gulliver. America's massive shift toward an information-based economy has been a mixed blessing. Because we have the most complex, multifaceted economy, we are a multifaceted target.

And our strategic vulnerability has risen hand-in-hand with our economic power. Like the Lilliputians, there are people who have used the principles of mathematics and science to master technology. They are so small in scale compared to the threats that we usually see that we have to strain our eyes just to identify them and figure out what they are doing. Gulliver did not win his freedom with a single act or weapon. He used a combination of things: sometimes he used his power, sometimes he used wit, and he learned from his experience how to deal with his adversaries.

We must do the same to both understand and defend our resources and position as a world leader in this new environment, and take into account the hidden dangers that lurk around every cyberspatial corner, across the oceans, down the street, in a corporate computer center, foreign intelligence service headquarters, or upstairs in a child's room.

In the opening paragraphs of this book, we refer to Information Warfare as a form of unconventional warfare. It is. The Information Warfare theater is the world, with no front lines, trenches, uniformed troops, or easy-to-see indications of an attack. While this emerging type of warfare is fought primarily with keyboards, computers, and networks, the traditional tools of war—guns, bombs, and troops—are also tools of the trade available for use in the Information War. Unfortunately, these tools of war are not aimed only against military targets, but strategic, civilian ones as well. After all, if one can decimate an enemy's military forces or cripple its economy or disrupt its citizen's sense of well-being and security, victory can be claimed, for the ultimate goal of any war is to deny the enemy their ability and/or will to fight. Information Warfare is just another way to make it happen, from anywhere, at any time.

As members of the most elite military and business community in the world, we must continue to harness this new technology and consider the strategic implications this new environment brings to our current doctrine. We must insure doctrine remains dynamic-ready to change and adapt to emerging trends in technology, adversarial tactics, and thought processes. As members of the most information-driven society in the world, we must guard against the misuse of information technology or its contents in a strategic information attack against not only the military forces, but those systems and infrastructures that so greatly affect civilians as well.

As the new millennium approaches, security professionals (like us) are responsible for safeguarding these systems and infrastructures have an interesting and challenging job. It falls to our small, elite cadre of information warriors to develop adequate responses to the ideology-, culturally-, and geographically-diverse technological threats of the future.

Onwards!

PART 2

Black Belt Knowledge

Techniques of the Information Warrior

A warrior's training is never complete. Learning by scroll is incomplete learning. True learning comes with each notch in the blade, gouge in the helm, and blood on the chain mail. Real knowledge comes to those who learn lessons from their failures and apply them toward their successes. The black belt of Experience should be worn with honor.

Sensei Sun

SECURITY ASSESSMENT PROCESSES

Security assessments need to support comprehensive evaluations of both the technical and non-technical measures implemented to protect information resources. The method chosen for a particular assessment must provide management with adequate information on which to base protection decisions. Based on the level of assurance intended, a security assessment commensurate with the assurance desired should be conducted.

In conducting an assessment, the following steps should be followed:

1) **Information Gathering:** This is the process of identifying all the assets associated with the information resource. It includes specific information necessary for conducting some of the various assessments. The information includes but is not limited to the following:
   a. Information Resource (Network) Topology and Inventory
      - Hardware
      - Software
      - Information/Data
      - Facilities
      - Support Systems
      - Personnel
      - Reputation

   b. IP addresses for scanning
   c. Phone numbers/exchanges to dial
   d. Information Resource (Network) Connectivity
   e. Identification of critical and sensitive resources
2) **System requirements examination:** This is a review of the requirements found in all applicable Laws, Rules, Executive Orders, organizational policies and procedures.
3) **Paper Penetration Analysis (Vulnerability Hypothesis):** This is an analysis of current known vulnerabilities associated with the information resource. This information is acquired from various Web sources or contacts (hacker pages, CIAC, CERT, etc.).
4) **Security review with System Owner representatives:** This is the process of attaining formal approval to pursue vulnerability inspections on the information resource from the system owner/manager.

5) **Investigation and attack:** The selection and implementation of assessment methods.

6) **Report and Recommendations:** This process includes the following:
   a. Outbriefing
      - Identify glaring vulnerabilities
      - Provide lists of cracked passwords
   b. Immediate Needs Report
      - Work-arounds and system patches
   c. Final Report
      - Analyze data
      - Document Assessment
      - Recommend corrective actions

7) **Configuration Accounting and Recording:** This is the process of recording the information resource assessment information into a database/spreadsheet. This information will be used as the baseline during follow-up reviews and assessments.

The following matrix below contains a list of the some of the various types of assessment processes that can be used to assess information resources, a description of the process, and the product (results) of the completed process.

Assessment Process	Description	Product
Social Engineering Exercise	An attack based on deceiving users or administrators at the target site. Social engineering attacks are typically carried out by telephoning users or operators and pretending to be an authorized user to attempt to gain illicit access to the systems.	Provides a measurement of the level of user awareness associated with a specific area of concern such as passwords and promotes a general security mindset.
Penetration Analysis	A type of security testing in which testers attempt to circumvent the security features of a system in an effort to identify system weaknesses.	Identifies weaknesses in system administration practices and procedures, and the presence/absence of controls.

Assessment Process	Description	Product
Security Processes and Documentation Review	This is a review of the existing security policies, procedures and guidelines.	Identifies strengths and weaknesses in the security documentation and program development.
User Interviews	Interviews conducted with resource users to determine their level of understanding of the security culture and requirements in place.	Identifies the awareness level associated with the security program.
War dialer exercise	A cracking tool, a program that calls a given list or range of numbers and records those which answer with handshake tones (and so might be entry points to computer or telecommunications systems).	Identifies possible weaknesses that may be exploited by outside adversaries.
Password cracking	The use of tools to decrypt weak passwords.	Verifies the password strength for system/application access.
Historical Incident Analysis	Review of historical records associated with incidents.	Provides a baseline of possible areas of concern associated with the information resource.
Paper Penetration Analysis (Vulnerability Hypothesis)	Examination of hacker bulletin boards, CIAC, CERT, etc. for known resource vulnerabilities.	Identifies specific areas of concern associated with information resources. Also provides recommendations to mitigate the concern.
Independent Verification and Validation (IV&V)	The function of performing rigorous testing of the developed system to verify and validate functional logic and technical implementation.	Provides quality control for software system configuration through periodic audits of both application and system software against the baseline.
Network Scanning	The use of automated tools to scan information resources for predetermined known vulnerabilities.	Provides a detailed description of the configuration parameters of the information resource. Identifies possible weaknesses.
Automated Risk assessment	A study of the vulnerabilities, threats, likelihood, loss or impact, and theoretical effectiveness of security measures. The process of evaluating threats and vulnerabilities, known and postulated, to determine expected loss and establish the degree of acceptability to system operations.	Provides management with the information upon which protection decisions can be based. Identifies specific weaknesses and associated safeguards. Provides a cost/benefit analysis for implementing the identified safeguards.

Security Assessment Steps

The following sections provide general steps associated with each type of assessment process identified above. The steps are examples that can be tailored for the users specific information resource.

Poomse 1: Social Engineering Exercise
Poomse 2: Penetration Analysis
Poomse 3: Security Processes, Controls and Documentation Review
Poomse 4: User Interview
Poomse 5: War Dialer Exercise
Poomse 6: Password Cracking
Poomse 7: Historical Incident Analysis
Poomse 8: Paper Penetration Analysis
Poomse 9: Independent Verification and Validation
Poomse 10: Network Scanning
Poomse 11: Introduction to Disaster Recovery

POOMSE 1

Social Engineering Exercises

Over the Phone

- Acquire a list of phone numbers for users of the information resource.
- Contact random users.
- Pretend to be an authorized System Administrator or other technical staff member.

Sample script: "Hello, my name is Bob Smith, I'm with network control. We're attempting to rebuild our Netware access control database which got damaged during an upgrade last night. You may have noticed the network running slower than usual today. Did you? Anyway, I show your network login ID to be "Jdoe" with a password of "goUSA". Is this correct? We would like to get this resolved before calling the vendor and having to disrupt the network with major repairs. Can you help us?" *The overall tone of the conversation should be pleasant and encourage willing cooperation from the unsuspecting user. By "playing up" the "major downtime" scenario the user will most likely want to cooperate—lest THEY be the cause of the "major network downtime" for the organization. Upon receiving the user's password, the assessor will comply with the approved risk assessment procedures (e.g., recording the success in gaining the password for statistical purposes or actually logging in as the user).*

- Document the results of the assessment.

In-Person

- Preplan your actions (vendor names, attire, schedules, points of contact, responsibilities, and so forth)
- Contact random users.
- Pretend to be an authorized vendor representative.

Sample script: "Hello, my name is Bob Smith, I'm with CompanyX, filling in for Chris, the regular support person. ManufacturerY has issued a warning to us about the installed network card on your file server, it may crash unexpectedly. Have you or anyone else noticed any weird network problems? If yes, I will need to run a few tests on the card from the file server console. Can you escort me to the file server or find someone who can login with system administrative privileges? *Upon unrestricted and hopefully unmonitored access to the file server, the assessment person can place a small harmless text file in the system area of the file server indicating his presence, complete with date, time and contact information of the assessment person.*

- Document the results of the assessment.

Authors note: Obtaining the password to typical office systems running Windows-based systems is child's play for the moderate-to-very-skilled Social Engineer. One great way to "prove" that you're a genuine caller is to ask if their Windows-based system has crashed recently—which, to the shrewd assessor, "is a sure sign of trouble" to help elicit information from unsuspecting users! And, by asking questions like "has your machine crashed recently?" or "has this happened to you?" you open yourself up for the unsuspecting victim to ask YOU questions—after all, why would a "bad guy" be asking me questions? Wouldn't they simply DO something to me or TELL me something wrong? Social Engineers that operate in an interactive mode have a much greater success rate and a lower chance of detection by their victims. And remember, as criminals know, ***the average American is far too trusting!***

POOMSE 2

Penetration Analysis

Internal Penetration

- Attain a list of information resources. (servers and workstations)
- Attempt to login to each resource using known accounts. (default or user)

The following is a sample list of common default accounts:

```
root       admin   guest          demo
sysadmin   unix    daemon         sysbin
uucp       rje     administrator  supervisor
```

The assessor will type one of the above account names in at the login prompt, then when asked for a password, the assessor would enter the account name again for the password. The following is an example:

```
Login> guest
Enter password> guest
```

If an authorized user ID is known, the assessor can try entering it at the login prompt. When asked for a password, the assessor would have to enter logical guesses (user's first or last name, type of vehicle the user drives, the user's birthday, children's name, hobbies, etc.). ***Note: Because of operating system security settings, multiple unsuccessful attempts using a valid user ID can cause the system to lock or disable the user's account. Insure you do not disrupt normal operations during your assessment.***

Upon gaining access to the file server, the assessment person can place a small harmless text file in the system area of the file server indicating his presence, complete with date, time and contact information of the assessment person.

- Document the results of the assessment.

External Penetration

- Determine topography of the information resource. (phone, internet, modem, etc.)
- Obtain a list of IP addresses for the information resource. (firewall and routers)
- Utilize automated scanning tools and personal knowledge, experience, and initiative to identify targets for directed attack. *See Poomse 10 for steps in conducting a Network Scan.*
- Utilize automated scanning tools for assessing identified modem resources. *See Poomse 5 for steps in conducting a War Dialer Exercise.*

Upon gaining access to the network through the firewall or router, the assessment person would then follow the steps outlined in the above Internal Penetration Process.

- Document the results of the assessment.

POOMSE 3

Security Processes and Documentation Review

- Obtain a copy of all security-related documentation.
- Disaster Recovery Plans
- Contingency Plans
- Security Plans
- Organization Policies, Procedures, and Guidelines
- Governmental Regulations
- Review historical audit and review documentation for identified security concerns.
- Review previous external and internal audit reports (contractor, internal assessment reports, etc.)
- Review requirements with applicable users.
- Interview Users/System Administrators/Managers for awareness knowledge of security requirements. *See Poomse 4 for the steps associated in conducting User Interviews.*
- Document the results of the assessment.

POOMSE 4

Interview Users

- Preplan interview questions based on job function.
- Develop interview questions for information resource users
- Develop interview questions for system administrators
- Develop interview questions for system managers

Questions should be derived from a review of the Organization's Processes, Controls, and Documentation. See Poomse 3 for steps in conducting this review.

- Schedule interviews with users/system administrators, system managers.
- Interview personnel by asking the questions developed above (NOTE: LIMIT THE NUMBER OF QUESTIONS PER INTERVIEWEE TO NO MORE THAN 30.) *The overall tone of the conversation should be pleasant and encourage willing cooperation from the interviewee. Explain to the interviewee the purpose of the interview, the expected outcome, and the importance of their input in relation to the organization. Include a "free-form" area for comments or concerns.*
- Document the results of the assessment.

POOMSE 5

War Dialer Exercise

- Obtain the necessary automated tools to conduct the exercise. *The following are some recommended tools:*

Toneloc

Sample parameter settings:

```
c:\toneloc [log filename] /m:688-xxxx /r:0001-1000
```

This configuration will dial all 688 extensions from 688-0001 through 688-1000 and send the results to a log file of your choice [log filename].

Phonetag

This is a Windows-based product that provides easily-configurable screens.

Vexwar

This a DOS-based program that steps the user through the configuration setup.

- Obtain a list (range) of phone numbers/exchanges to dial.
- Configure the automated tool.
- Select the range of phone numbers to be dialed
- Select the START and STOP time for the scan
- Enter a log filename for the data to be collected in
- Enter any other parameters desired *(Assessor can read the read.me or other program files to determine what parameters to set.)*

- Initiate the exercise.
- Review the data log file for identified modem connections *(The log file will identify a modem connection by displaying a login prompt for an identified phone number.)*
- Dial the identified connections.
- Follow the steps identified *for an Internal Penetration Analysis in Poomse 2 after a connection is made.*
- Document the results of the assessment.

POOMSE 6

Password Cracking

- Obtain the necessary automated tools to conduct the exercise.

SUGGESTED PASSWORD CRACKERS	
File Name	**Description**
XIT 2.0	DOS Based cracker with Source Code (Borland C)
PCUPC	The Personal Computer UNIX Password Cracker (PCUPC) is a program which effectively encrypts passwords from a list of words, comparing them to any UNIX etc/password file's encrypted passwords; matching passwords are recorded for future use.
Killer Cracker v.9.5	Disk Jockey modified Killer Cracker v9.5—One of the fastest crackers around.
Glide	Cracks the Win95 password file
Crack v4.1f	Password cracker for OS/2
Brute Force	A brute force approach to hacking UNIX password
Cracker Jack v.4.1	A UNIX password checker/cracker
NewPW Cracker	UNIX-based password cracking program
JackAss v1.1	UNIX Based (to crack UNIX Password Files)
Killer Cracker v9.11	UNIX-based FAST password-cracking program
Hellfire Cracker V1.3	Very slow but effective UNIX-based password cracking program
Lard v2.0	UNIX-based password cracking program
crack-2a.tgz	UNIX Password Cracker 2.0(a) by Scooter Corp. (Comes with crack dictionary).
Crack v5.0	A pretty decent password cracker, this is the one to use for Linux
Novelbfh	A Novell password cracker
L0phtcrack	Windows NT password cracker
Phcrack	Windows NT password cracker
Vines Security Authenticator	Banyan Vines password cracker

- Obtain authorization from the System Owner to conduct the cracking exercise.
- Obtain the necessary access rights needed to run the password cracking program from the system administrator.
- Configure the software.
- Select the appropriate dictionary files to be used
- Initiate the exercise. *(NOTE: Conduct the exercise on a copy of the original password file in a temporary working directory.)*
- Document the results of the assessment.

POOMSE 7

Historical Incident Analysis

- Obtain a copy of all security-related Incident documentation.
- Review the documentation for security concerns.
- Compile a list of follow-up security concerns
- Conduct follow-up reviews with applicable users.
- Interview Users/System Administrators/Managers to identify what actions were taken in association with incident report concerns. *See Poomse 4 for the steps associated in conducting User Interviews.*
- Document the results of the assessment.

Paper Penetration Analysis
(Vulnerability Hypothesis)

- Conduct a search of Hacker Web Sites on the Internet.
- Compile a list of vulnerabilities identified for the information resource being reviewed.
- Search for keywords that might be associated with your organization.
- Conduct a search on the CERT Web Site on the Internet <*www.cert.org*>.
- Compile a list of vulnerabilities and patches from advisories applicable to the information resource.
- Conduct a search on the CIAC Web Site on the Internet <*www.ciac.org*>.
- Compile a list of vulnerabilities and patches from bulletins applicable to the information resource.
- Conduct a search on the Vendor's Web Site and other security sites on the Internet for the information resource.
- Compile a list of vulnerabilities and patches applicable to the information resource.
- Conduct an Interview with the System Administrator/Manager *(See Poomse 4 for the steps associated in conducting User Interviews.)*
- Review the identified vulnerabilities with the System Administrator/Manager.
- Verify that all patches (fixes) for applicable vulnerabilities have been implemented.
- Document the results of the assessment.

Independent Verification and Validation (IV&V)

- Review all documentation used to establish the baseline requirements associated with the information resource during its development and implementation stages.
- Compile a baseline of requirements for the information resource based on the documentation.
- Obtain authorization from the system owner/manager to review the baseline requirements against the current configuration settings.
- Conduct a review of the information resource with the system administrator/owner/manager.
- Review the current configuration settings.
- Document the differences between the initial baseline settings and the current settings.
- Validate justifications for the differences from the system owner/manager.
- Document the justified differences as the new baseline.
- Document the results of the assessment.

Network Scanning

- Obtain the necessary automated tools to conduct the exercise. *The following are some suggested tools:*

  SATAN (Security Administrator Tool for Analyzing Networks)

  SATAN is a tool to help systems administrators. It recognizes several common networking-related security problems, and reports the problems without actually exploiting them.

  ISS (Internet Security Scanner)

  The Internet Scanner set of tools perform scheduled and selective probes of your network's communication services, operating systems, key applications, and routers in search of those vulnerabilities most often used by unscrupulous threats to probe, investigate, and attack your network. Internet Scanner then analyzes your vulnerability conditions and provides a series of corrective action, trends analysis, conditional, and configuration reports and data sets.

  NetRecon

  NetRecon probes networks and network resources and displays vulnerabilities as they are detected and quickly perform deeper probes. This makes it easy to understand the ramifications of security problems so you know which ones are the most important.

- Obtain authorization from the System Owner to conduct the scanning exercise.
- Obtain the necessary access rights needed to run the network scanning program from the system administrator.
- Configure the software.
- Set the boundaries for the scan

  Which network resources to scan

  Servers
  Domains
  Workstation

  Select the desire scanning parameters

  Specific vulnerabilities

  Requirement settings (password length and expiration, audit logging, etc.)

- Initiate the exercise.
- Verify the network topology
- Identify resources with know vulnerabilities
- Document the results of the assessment.

Introduction to Disaster Recovery

Any discussion of risk assessment programs would be incomplete without at least touching on disaster recovery or "business continuity" planning. A risk assessment is the first step in developing plans and procedures for the safeguarding of organizational assets not only prior to an incident, but also to act upon in the event of an incident against the assets identified in the course of the risk assessment.

A computer security contingency is an event with the potential to disrupt computer operations, thereby disrupting critical mission and business functions. Such an event could be a power outage, hardware failure, fire, country uprising, or storm. If the event is very destructive, it is often called a disaster.

To avert potential contingencies and disasters or minimize the damage they cause, organizations can take steps early to control the event. Generally called contingency planning, this activity is closely related to incident handling, which primarily addresses malicious technical threats such as hackers and viruses.

Contingency Planning

Contingency planning involves more than planning for a move off-site after a disaster destroys a data center. It also addresses how to keep an organization's critical functions operating in the event of disruptions, both large and small. This broader perspective on contingency planning is based on the distribution of computer support throughout an organization. The contingency planning process involves the following steps:

- Identifying the mission—or business-critical functions,
- Identifying the resources that support the critical functions,
- Anticipating potential contingencies or disasters,
- Selecting contingency planning strategies,
- Implementing the contingency strategies, and
- Testing and revising the strategy.

Identifying the Mission—or Business-Critical Functions

Protecting the continuity of an organization's mission or business is very difficult if it is not clearly identified. Managers need to understand the organization from a point of view that usually extends beyond the area they control. The definition of an organization's critical mission or business functions is often called a business plan.

Since the development of a business plan will be used to support contingency planning, it is necessary not only to identify critical missions and businesses, but also to set priorities for them. A fully redundant capability for each function is prohibitively expensive for most organizations. In the event of a disaster, certain functions will not be performed. If appropriate priorities have been set (and approved by senior management), it could mean the difference in the organization's ability to survive a disaster.

Identifying the Resources That Support Critical Functions

After identifying critical missions and business functions, it is necessary to identify the supporting resources, the time frames in which each resource is used (e.g., is the resource needed constantly or only at the end of the month?), and the effect on the mission or business of the unavailability of the resource. In identifying resources, a traditional problem has been that different managers oversee different resources. They may not realize how resources interact to support the organization's mission or business. Many of these resources are not computer resources. Contingency planning should address all the resources needed to perform a function, regardless whether they directly relate to a computer.

The analysis of needed resources should be conducted by those who understand how the function is performed and the dependencies of various resources on other resources and other critical relationships. This will allow an organization to assign priorities to resources since not all elements of all resources are crucial to the critical functions.

Human Resources

People are perhaps an organization's most obvious resource. Some functions require the effort of specific individuals, some require specialized expertise, and some only require individuals who can be trained to perform a specific task. Within the information technology field, human resources include both operators (such as technicians or system programmers) and users (such as data entry clerks or information analysts).

Processing Capability

Traditionally contingency planning has focused on processing power (i.e., if the data center is down, how can applications dependent on it continue to be processed?). Although the need for data center backup remains vital, today's other processing alternatives are also important. Local area networks (LANs), minicomputers, workstations, and personal computers in all forms of centralized and distributed processing may be performing critical tasks.

Automated Applications and Data

Computer systems run applications that process data. Without current electronic versions of both applications and data, computerized processing may not be possible. If the processing is being performed on alternate hardware, the applications must be compatible with the alternate hardware, operating systems and other software (including version and configuration), and numerous other technical factors. Because of the complexity, it is normally necessary to periodically verify compatibility. (See Testing and Revising.)

Computer-Based Services

An organization uses many different kinds of computer-based services to perform its functions. The two most important are normally communications services and information services. Communications can be further categorized as data and voice communications; however, in many organizations these are managed by the same service. Information services include any source of information outside of the organization. Many of these sources are becoming automated, including on-line government and private databases, news services, and bulletin boards.

Physical Infrastructure

For people to work effectively, they need a safe working environment and appropriate equipment and utilities. This can include office space, heating, cooling, venting, power, water, sewage, other utilities, desks, telephones, fax machines, personal computers, terminals, courier services, file cabinets, and many other items. In addition, computers also need space and utilities, such as electricity. Electronic and paper media used to store applications and data also have physical requirements.

Documents and Papers

Many functions rely on vital records and various documents, papers, or forms. These records could be important because of a legal need (such as being able to produce a signed copy of a loan) or because they are the only record of the information. Records can be maintained on paper, microfiche, microfilm, magnetic media, or optical disk.

Anticipating Potential Contingencies or Disasters

Although it is impossible to think of all the things that can go wrong, the next step is to identify a likely range of problems. The development of scenarios will help an organization develop a plan to address the wide range of things that can go wrong.

Scenarios should include small and large contingencies. While some general classes of contingency scenarios are obvious, imagination and creativity, as well as research, can point to other possible, but less obvious, contingencies. The contingency scenarios should address each of the resources described above. The following are examples of some of the types of questions that contingency scenarios may address:

Human Resources

Can people get to work? Are key personnel willing to cross a picket line or riot zone? Are there critical skills and knowledge possessed by one person? Can people easily get to an alternative site?

Processing Capability

Are the computers harmed? What happens if some of the computers or networks are inoperable, but not all?

Automated Applications and Data

Has data integrity been affected? Is an application sabotaged? Can an application run on a different processing platform?

Computer-Based Services

Can the computers communicate? To where? Can people communicate? Are information services down? For how long? Do we have an alternate communications provider?

Physical Infrastructure

Do people have a place to sit? Do they have equipment to do their jobs? Can they occupy the building?

Documents and Papers

Can needed records be found? Are they readable?

Selecting Contingency Planning Strategies

The next step is to plan how to recover needed resources. In evaluating alternatives, it is necessary to consider what controls are in place to prevent and minimize contingencies. Since no set of controls can cost-effectively prevent all contingencies, it is necessary to coordinate prevention and recovery efforts.

A contingency planning strategy normally consists of three parts: emergency response, recovery, and resumption. Emergency response encompasses the initial actions taken to protect lives and limit damage. Recovery refers to the steps that are taken to continue support for critical functions. Resumption is the return to normal operations. The relationship between recovery and resumption is important. The longer it takes to resume normal operations, the longer the organization will have to operate in the recovery mode.

The selection of a strategy needs to be based on practical considerations, including feasibility and cost. The different categories of resources should each be considered. Risk assessment can be used to help estimate the cost of options to decide on an optimal strategy. For example, is it more expensive to purchase and maintain a generator or to move processing to an alternate site, considering the likelihood of losing electrical power for various lengths of time? Are the consequences of a loss of computer-related resources sufficiently high to warrant the cost of various recovery strategies? The risk assessment should focus on areas where it is not clear which strategy is the best.

In developing contingency planning strategies, there are many factors to consider in addressing each of the resources that support critical functions.

Human Resources

To ensure an organization has access to workers with the right skills and knowledge, training and documentation of knowledge are needed. During a major contingency, people will be under significant stress and may panic. If the contingency is a regional disaster, their first concerns will probably be their family and property. In addition, many people will be either unwilling or unable to come to work. Additional hiring or temporary services can be used. The use of additional personnel may introduce security vulnerabilities.

Contingency planning, especially for emergency response, normally places the highest emphasis on the protection of human life.

Processing Capability

Strategies for processing capability are normally grouped into five categories: hot site; cold site; redundancy; reciprocal agreements; and hybrids. These terms originated with recovery strategies for data centers but can be applied to other platforms.

- **Hot site**—A building already equipped with processing capability and other services.

- **Cold site**—A building for housing processors that can be easily adapted for use.
- **Redundant site**—A site equipped and configured exactly like the primary site. (Some organizations plan on having reduced processing capability after a disaster and use partial redundancy. The stocking of spare personal computers or LAN servers also provides some redundancy.)
- **Reciprocal agreement**—An agreement that allows two organizations to back each other up. (While this approach often sounds desirable, contingency planning experts note that this alternative has the greatest chance of failure due to problems keeping agreements and plans up-to-date as systems and personnel change.)
- **Hybrids**—Any combinations of the above such as using having a hot site as a backup in case a redundant or reciprocal agreement site is damaged by a separate contingency.

Recovery may include several stages, perhaps marked by increasing availability of processing capability. Resumption planning may include contracts or the ability to place contracts to replace equipment.

Automated Applications and Data

Normally, the primary contingency strategy for applications and data is regular backup and secure offsite storage. Important decisions to be addressed include how often the backup is performed, how often it is stored offsite, and how it is transported (to storage, to an alternate processing site, or to support the resumption of normal operations).

Computer-Based Services

Service providers may offer contingency services. Voice communications carriers often can reroute calls (transparently to the user) to a new location. Data communications carriers can also reroute traffic. Hot sites are usually capable of receiving data and voice communications. If one service provider is down, it may be possible to use another. However, the type of communications carrier lost, either local or long distance, is important. Local voice service may be carried on cellular. Local data communications, especially for large volumes, is normally more difficult. In addition, resuming normal operations may require another rerouting of communications services.

Physical Infrastructure

Hot sites and cold sites may also offer office space in addition to processing capability support. Other types of contractual arrangements can be made for office space, security services, furniture, and more in the event of a contingency. If the contingency plan calls for moving offsite, procedures need to be developed to ensure a smooth transition back to the primary operating facility or to a new facility. Protection of the physical infrastructure is normally an important part of the emergency response plan, such as use of fire extinguishers or protecting equipment from water damage.

Documents and Papers

The primary contingency strategy is usually backup onto magnetic, optical, microfiche, paper, or other medium and offsite storage. Paper documents are generally harder to backup than electronic ones. A supply of forms and other needed papers can be stored offsite.

Implementing the Contingency Strategies

Once the contingency planning strategies have been selected, it is necessary to make appropriate preparations, document the strategies, and train employees. Many of these tasks are ongoing.

Much preparation is needed to implement the strategies for protecting critical functions and their supporting resources. For example, one common preparation is to establish procedures for backing up files and applications. Another is to establish contracts and agreements, if the contingency strategy calls for them. Existing service contracts may need to be renegotiated to add contingency services. Another preparation may be to purchase equipment, especially to support a redundant capability.

It is important to keep preparations, including documentation, up-to-date. Computer systems change rapidly and so should backup services and redundant equipment. Contracts and agreements may also need to reflect the changes. If additional equipment is needed, it must be maintained and periodically replaced when it is no longer dependable or no longer fits the organization's architecture.

Preparation should also include formally designating people who are responsible for various tasks in the event of a contingency. These people are often referred to as the contingency response team. This team is often composed of people who were a part of the contingency planning team.

There are many important implementation issues for an organization. Two of the most important are how many plans should be developed and who prepares each plan. Both of these questions revolve around the organization's overall strategy for contingency planning. The answers should be documented in organization policy and procedures.

How Many Plans?

Some organizations have just one plan for the entire organization, and others have a plan for every distinct computer system, application, or other resource. Other approaches recommend a plan for each business or mission function, with separate plans, as needed, for critical resources.

The answer to the question, therefore, depends upon the unique circumstances for each organization. But it is critical to coordinate between resource managers and functional managers who are responsible for the mission or business.

Who Prepares the Plan?

If an organization decides on a centralized approach to contingency planning, it may be best to name a contingency planning coordinator. The coordinator prepares the plans in cooperation with various functional and resource managers. Some organizations place responsibility directly with the functional and resource managers.

Documenting

The contingency plan needs to be written, kept up-to-date as the system and other factors change, and stored in a safe place. A written plan is critical during a contingency, especially if the person who developed the plan is unavailable. It should clearly state in simple language the sequence of tasks to be performed in the event of a contingency so that someone with minimal knowledge could immediately begin to execute the plan. It is generally helpful to store up-to-date copies of the contingency plan in several locations, including any off-site locations, such as alternate processing sites or backup data storage facilities.

Training

All personnel should be trained in their contingency-related duties. New personnel should be trained as they join the organization, refresher training may be needed, and personnel will need to practice their skills.

Training is particularly important for effective employee response during emergencies. There is *no time* to check a manual to determine correct procedures if there is a fire, earthquake, or terrorist explosion. Depending on the nature of the emergency, there may or may not be time to protect equipment and other assets. Practice is necessary in order to react correctly, especially when human safety is involved. The ancient adage of "Train, Train, Train" comes to mind, to the point where the individuals know what to do on *instinct* versus *insight.*

Testing and Revising

A contingency plan should be tested periodically because there will undoubtedly be flaws in the plan and in its implementation. The plan will become dated as time passes and as the resources used to support critical functions change. Responsibility for keeping the contingency plan current should be specifically assigned. The extent and frequency of testing will vary between organizations and among systems. There are several types of testing, including reviews, analysis, and simulations of disasters.

A review can be a simple test to check the accuracy of contingency plan documentation. For instance, a reviewer could check if individuals listed are still in the organization and still have the responsibilities that caused them to be included in the plan. This test can check home and work telephone numbers, organizational codes, and building and room numbers. The review can determine if files can be restored from backup tapes or if employees know emergency procedures.

An analysis may be performed on the entire plan or portions of it, such as emergency response procedures. It is beneficial if the analysis is performed by someone who did not help develop the contingency plan but has a good working knowledge of the critical function and supporting resources. The analyst(s) may mentally follow the strategies in the contingency plan, looking for flaws in the logic or process used by the plan's developers. The analyst may also interview functional managers, resource managers, and their staff to uncover missing or unworkable pieces of the plan.

Organizations may also arrange disaster simulations. These tests provide valuable information about flaws in the contingency plan and provide practice for a real emergency. While they can be expensive, these tests can also provide critical information that can be used to ensure the continuity of important functions. In general, the more critical the functions and the resources addressed in the contingency plan, the more cost-beneficial it is to perform a disaster simulation.

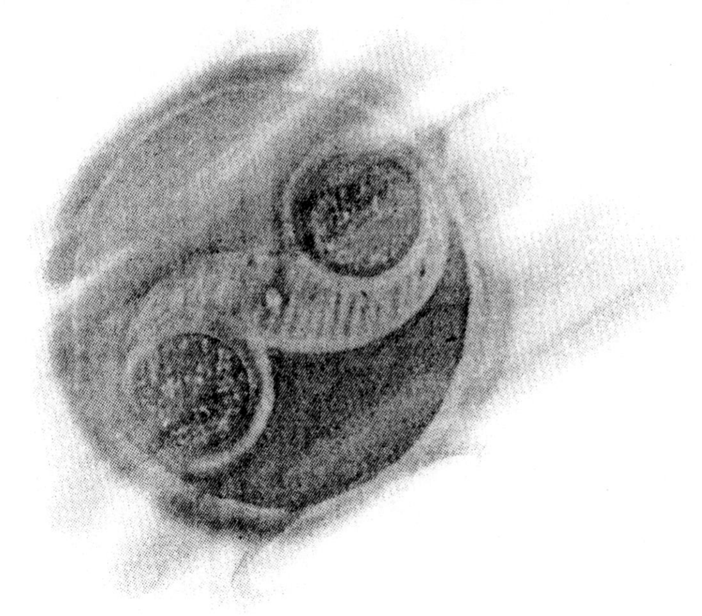

The only bad actions are uninformed ones.

Appendix A—List of Computer Emergency Response Organizations

Appendix B—Relevant US Government, Commercial, and Non-Profit Organizations

Appendix C—Other Resources

Appendix D—Glossary of InfoWar Terms

APPENDIX A

Partial Listing of Computer Emergency Response Teams

United States Government and Military Teams

AFCERT (U.S. Air Force CERT)

E-mail: <afcert@afcert.csap.af.mil>
Telephone: +1 210-977-3157
Telephone: +1 800-854-0187
Fax: +1 210-977-3632

ASSIST (US. Department of Defense Automated Systems Security Incident Support Team)

E-mail: <assist@assist.mil>
Telephone: +1 800-357-4231
Fax: +1 703-607-4735

CIAC (US. Department of Energy's Computer Incident Advisory Capability)

E-mail: <ciac@llnl.gov>
Telephone: +1 925-422-8193, 24/7
Fax: +1 925-423-8002

NIST/CSRC

E-mail: <first-team@csmes.ncsl.nist.gov>
Telephone: +1 301-975-3359
Fax: +1 301-948-0279

NAVCIRT (Naval Computer Incident Response Team)

E-mail: <navcirt@fiwc.navy.mil>
Telephone: +1 757-417-4024
Telephone: +1 888-NAVCIRT (628-2478)
Telephone: +1 800-628-8893

NIH CERT (US. National Institutes of Health)

E-mail: <Kevin_Haney@nih.gov>
Telephone: +1 301 402-1812
Telephone: +1 301 594-3278 (emerg.)
Fax: +1 301 402-1620

NIST/CSRC

E-mail: <first-team@csmes.ncsl.nist.gov>
Telephone: +1 301-975-3359
Fax: +1 301-948-0279

SSACERT (U.S. Social Security Administration)

E-mail: <ssacert@ssa.gov>
Telephone: +1 410 966-9075 or +1 410 965-6950
Fax: +1 410 966-6230

USHCERT (US. House of Representatives Computer Emergency Response Team)

E-mail: <security@mail.house.gov>
Telephone: +1 202-226-4316
Pager: +1 800-SKY-8888 pin 1973770
Fax: +1 202-225-0368

International Response Teams

AUSCERT (Australian Computer Emergency Response Team)

E-mail: <auscert@auscert.org.au>
Telephone: +61 7-3365-4417, 24/7
Fax: +61 7-3365-4477

CARNet-CERT (Croatia)

E-mail: <c-cert@carnet.hr>
Telephone: +385 1-45-94-337

CCTA (UK Government Agencies)

E-mail: <cbaxter.ccta.esb@gnet.gov.uk>
Telephone: +44 0171-824-4101/2
Fax: +44 0171-305-3178

CERT® Coordination Center (CERT/CC)—"Official" CERT

E-mail: <cert@cert.org>
Telephone: +1 412-268-7090
Fax: +1 412-268-6989

CERT-IT (Computer Emergency Response Team Italiano)

E-mail: <cert-it@dsi.unimi.it>
Telephone: +39 2-55006-300, +39 2-55006-387
Fax: +39 2-55006-388

CERTCC-KR (Computer Emergency Response Team Korea)

E-mail: <cert@certcc.or.kr>
Telephone: +82-2-3488-4119
Fax: +82-2-3488-4129
Cellular Phone : +82-18-312-4119
Pager: +82-15-993-4571

CERT-NL (Constituency: Dutch academic, educational and research network)

E-mail: <cert-nl@surfnet.nl>
Telephone: +31 30-2305305
Fax: +31 30-2305329
URGENT CSIR CALLS: +31 6 52 87 92 82 (AT ALL TIMES)

DANTE (Delivery of Advanced Network Technology to Europe Ltd.)

E-mail: <SafeFLOW@dante.org.uk>
Telephone: +44 1223-302-992
Fax: +44 1223-303-005

DFN-CERT (Germany CERT)

E-mail: <dfncert@cert.dfn.de>
Telephone: +49 40-5494-2262
Fax: +49 40-5494-2241

DK-CERT (Danish Computer Emergency Response Team)

E-mail: <cert@cert.dk>
Telephone: +45 35 87 88 89
Fax: +45 35 87 88 90

EuroCERT

E-mail: <eurocert@eurocert.net>
Telephone: +44 1235 822 240
Fax: +44 1235 822 398

IRIS-CERT (Spain)

E-mail: <cert@rediris.es>
Telephone: +34 91-585-4992 or +34 91-585-5150
(9am–6pm, MET (GMT+01))
Fax: +34 91-585-5146

JANET-CERT (Constituency: All UK organizations connected to JANET network)

E-mail: <cert@cert.ja.net>
Telephone: +44 1235-822-340
Fax: +44 1235-822-398
Membership Type: Full member

MxCERT (Mexican CERT)

E-mail: <mxcert@mxcert.org.mx>
Telephone: +52 8-328-4088
Fax: +52 8-328-4129

SWITCH-CERT (Swiss Academic and Research Network CERT)

E-mail: <cert-staff@switch.ch>
Telephone: +41 1-268-1518
Fax: +41 1-268-1568

Commercial Response Teams

Note: Only organizations with response teams that include customers, other external clients, or the Internet community are listed below. Organizations with teams exclusively for internal response support are not included in this list.

Apple Computer

E-mail: <first-team@apple.com>
Telephone: +1 408-974-6985
Fax: +1 408-974-1560

AT&T

E-mail: <first-team@inetmail.att.net>
Telephone: +1 732 576 7153
Fax: +1 732 576 4473

Cisco Product Security Incident Response Team (Cisco-PSIRT)

E-mail: <first-team@cisco.com>
Telephone: +1 408 526-8212
(approx. 11:00–19:00 GMT-0800, M-F)
Telephone: +1 800 553-6387
(Emergencies only, ask for PSIRT/PSIAP access)
Fax: +1 408 527-2206
(unsecured FAX, voice call followup recommended)

FreeBSD, Inc.

E-mail: <security-officer@freebsd.org>
Telephone: +31 40 2127794

Goldman, Sachs and Company

E-mail: <shabbir.safdar@gs.com>
Telephone: +1 212-357-1880
Pager: +1 917-978-8430

Hewlett-Packard Company

E-mail: <security-alert@hp.com>

IBM-ERS (IBM Emergency Response Service)

E-mail: <ers@vnet.ibm.com>
Telephone: +1 914 759-4452
(8am–5pm, EST/EDT (GMT-5/GMT-4))
Telephone: +1 914 364-0199 (after hours)
Fax: +1 914-759-4326
Pager: +1 800-759-8352, PIN 1081136
(alphanumeric, two-way)
Pager: 1081136@skytel.com

MCI

E-mail: <Kevin.J.McMahon@mci.com>
Telephone: +1 703-506-6294
Pager: +1 800-SKY-8888 pin 216-2056
Fax: +1 703-506-6281

Micro-BIT Virus Center

E-mail: <ry15@uni-karlsruhe.de>
Telephone: +49 721 9620122
Telephone: +49 171 5251685 (emerg.)
Fax: +49 721 9620199

SGI (Silicon Graphics Inc.)

E-mail: <security-alert@sgi.com>
Telephone: +1 415-933-4997
Fax: +1 415-961-6502

Sprint

E-mail: <security@sprint.net>

Telephone: +1 800-572-8256

Fax: +1 703-478-5468

SUN Microsystems, Inc.

E-mail: <chok@barrios.eng.sun.com>

Telephone: +1 415-786-4420

Fax: +1 415-786-7994

Relevant Organizations

US Government

President's Commission on Critical Infrastructure Protection (PCCIP)
http://www.pccip.gov/

National Infrastructure Protection Center (NIPC)
@Federal Bureau of Investigation
http://www.fbi.gov/nipc/index.htm

Critical Infrastructure Protection Office (CIAO)
@ Department of Commerce
http://www.ciao.gov/

FBI Awareness of National Security Issues and Response (ANSIR) Program
http://www.fbi.gov/ansir/ansir.htm

Defense Information Systems Agency
http://www.disa.mil/

National Security Agency
http://www.nsa.gov:8080/

Department of Justice Computer Crimes and Intellectual Property Section
http://www.usdoj.gov/criminal/cybercrime/index.html

Computer Security Clearinghouse
@National Institute of Standards & Technology
http://www.csrl.nist.gov/

Commercial/Academic

Info-War
The Net's leading Information Warfare site
http://www.infowar.com/

The Art of Information Warfare
Security Education and Outreach Developers
http://www.taoiw.org/

Computer Security Institute
Computer Security Conferences and Training
http://www.gocsi.com/

New Technologies, Inc.
Computer Forensics, Consulting, Law Enforcement Training
http://www.secure-data.com/

Network Associates, Inc.
Makers of PGP, Virus Scan, and other security software
http://www.nai.com/

Internet Security Systems
Makers of the ISS Network Scanning Suite and Real Secure Network Intrusion Detectors.
http://www.iss.net/

Security Dynamics
Makers of authentication tokens, scanners, and encryption tools
http://www.securitydynamics.com/

The Operations Security Professionals Society
Association of security and counterintelligence professionals; annual training; certifications.
http://www.opsec.org/

High Technology Crime Investigations Assn.
Networking security, law enforcement, and investigators
http://htcia.org/

International Information Systems Security Certification Consortium
Professional certifications; training; networking
http://www.isc2.org/

American Society for Industrial Security
Training, networking, certifications
http://www.asisonline.org/

Other Resources

Security And Private Organizations

Information Systems Security Association (ISSA)
401 N. Michigan Ave.
Chicago, Illinois 60611
312-644-6610

MIS Training Institute
498 Concord St.
Framingham, Mass. 01701
508-879-7999
www.misti.com

Electronic Privacy Information Center
666 Pennsylvania Ave.
Washington, D.C. 20003
202-544-9240
www.epic.org

Computer Professionals for Social Responsibility
666 Pennsylvania Ave.
Washington, D.C. 20003
202-544-9240
www.cpsr.org

New Technologies Inc. Computer Forensics Training
2075 Northeast Division
Gresham, OR 97030
503-666-6599
www.secure-data.com

Electronic Frontier Foundation

666 Pennsylvania Ave.
Washington, D.C. 20003
202-544-9237
www.eff.org

Security and Privacy Publications

Information Security News

498 Concord St.
Framingham, Mass. 01701
508-879-7999

Privacy Journal

PO Box 28577
Providence, Rhode Island 02908
401-274-7861

Virus Bulletin

21 The Quadrant
Abingdon Science Park
Abingdon, Oxfordshire
0X14 3YS
United Kingdom

Data Pro Research

600 Delran Parkway
Delran, New Jersey 08075
800-328-2772

Security Management

1655 North Fort Myer Drive
Arlington, Virginia 22209
703-522-5800

Cryptologia

17 Alfred Road West
Merrick, New York 11566
516-378-0263

Computer Security Digest

150 N. Main
Plymouth, Michigan 48170
313-459-8787

Internet World

11 Ferry Lane West
Westport, Conn. 06880
203-226-6967

Wired

544 Second St.
San Francisco, California 94107
415-904-0660

Hacker Meetings

PumpCon—Philadelphia, October. Raided for two straight years. Information circulated on the net a few months prior to the event.

SummerCon—The famous yearly private hacker gathering. Invitation only.

HoHo Con—Texas, December. Hence the name.

DefCon—Las Vegas, Summer. More organized than the rest, the "granddaddy" of hacker cons.

2600 Hacker Meetings—First Friday of every month at worldwide locations. Check out the official website (www.2600.com) for places near you. Everyone is invited— hackers to professionals to law enforcement to those without a clue.

Black Hat Briefings—Annual technical computer security conferences.

Electronic Mail Distribution Lists

Alert

To join, send e-mail to majordomo@iss.net and, in the text of your message (not the subject line), write:

subscribe alert

To remove, send e-mail to majordomo@iss.net and, in the text of your message (not the subject line), write:

unsubscribe alert

This is a moderated list in the effort to keep the noise to a minimal and provide quality security information. The Alert will be covering the following topics:

Security Product Announcements
Updates to Security Products
New Vulnerabilities found
New Security Frequently Asked Question files.
New Intruder Techniques and Awareness

Bugtraq

To join, send e-mail to LISTSERV@NETSPACE.ORG and, in the text of your message (not the subject line), write:

subscribe bugtraq

This list is for *detailed* discussion of UNIX security holes: what they are, how to exploit, and what to do to fix them.

This list is not intended to be about cracking systems or exploiting their vulnerabilities. It is about defining, recognizing, and preventing use of security holes and risks.

Please refrain from posting one-line messages or messages that do not contain any substance that can relate to this list's charter.

Please follow the below guidelines on what kind of information should be posted to the Bugtraq list:

Information on Unix related security holes/backdoors (past and present)

Exploit programs, scripts or detailed processes about the above

Patches, workarounds, fixes

Announcements, advisories or warnings

Ideas, future plans or current works dealing with Unix security

Information material regarding vendor contacts and procedures

Individual experiences in dealing with above vendors or security organizations

Incident advisories or informational reporting

COAST Security Archive

To join, send e-mail to coast-request@cs.purdue.edu and, in the text of your message (not the subject line), write:

subscribe coast

Computer Privacy Digest

To join, send e-mail to comp-privacy-request@uwm.edu and, in the text of your message (not the subject line), write:

subscribe cpd

The Computer PRIVACY Digest (CPD) (formerly the Telecom Privacy digest) is run by Leonard P. Levine. It is transferred to the USENET newsgroup comp.society.privacy. It is a relatively open (i.e., less tightly moderated) forum, and was established to provide a forum for discussion on the effect of technology on privacy. All too often technology is way ahead of the law and society as it presents us with new devices and applications. Technology can enhance and detract from privacy.

Computer Underground Digest

To join, send e-mail to CU-DIGEST-REQUEST@WEBER.UCSD.EDU and, in the text of your message (not the subject line), write:

sub cudigest

CuD is available as a Usenet newsgroup: comp.society.cu-digest. Covers many issues of the computer underground.

Cypherpunks

To join, send e-mail to majordomo@toad.com and, in the text of your message (not the subject line), write:

subscribe cypherpunks-unedited

The cypherpunks list is a forum for discussing personal defenses for privacy in the digital domain. It is a high volume mailing list.

Cypherpunks Announce

To join, send e-mail to majordomo@toad.com and, in the text of your message (not the subject line), write:

subscribe cypherpunks-announce

There is an announcements list which is moderated and has low volume. Announcements for physical cypherpunks meetings, new software and important developments will be posted there.

Euro Firewalls

To join, send e-mail to majordomo@gbnet.net and, in the text of your message (not the subject line), write:

subscribe firewalls-uk e-mail-addr

European-oriented firewall list.

Firewalls

To join, send e-mail to majordomo@lists.gnac.net and, in the text of your message (not the subject line), write:

subscribe firewalls

Useful information regarding firewalls and how to implement them for security.

This list is for discussions of Internet "firewall" security systems and related issues. It is an outgrowth of the Firewalls BOF session at the Third UNIX Security Symposium in Baltimore on September 15, 1992.

INFSEC-L Information Systems Security Forum

To join, send e-mail to listserv@etsuadmn.etsu.edu and, in the text of your message (not the subject line), write:

sub infsec-l your-name

INFSEC-L is for discussions of information systems security and related issues. Discussions are not moderated. Thus, all messages sent to the list are immediately distributed to members of the list. The discussion list is an outgrowth of the "Technology for the Information Security '94: Managing Risk" at Galveston, TX (December 5–8, 1994). The main objective of the list is to foster open and constructive communication among information systems security and auditing professionals in government, industry, and academic institutions. Initial subscriptions are screened by the list moderator to ensure that only appropriate professionals are subscribed.

Intrusion Detection Systems

To join, send e-mail to majordomo@uow.edu.au with the following in the body of the message:

subscribe ids

The list is a forum for discussions on topics related to development of intrusion detection systems. Possible topics include:

techniques used to detect intruders in computer systems and computer

networks

audit collection/filtering

subject profiling

knowledge based expert systems

fuzzy logic systems

neural networks

methods used by intruders (known intrusion scenarios)

CERT advisories

scripts and tools used by hackers

computer system policies

universal intrusion detection system

NTBugtraq

NTBugtraq is a mailing list for the discussion of security exploits and security bugs in Windows NT and its related applications.

To join, send e-mail to listserv@listserv.ntbugtraq.com and, in the text of your message (not the subject line), write:

subscribe ntbugtraq

To remove, send e-mail to listserv@listserv.ntbugtraq.com and, in the text of your message (not the subject line), write:

unsubscribe ntbugtraq

NT Security

To join, send e-mail to majordomo@iss.net and, in the text of your message (not the subject line), write:

subscribe ntsecurity

To remove, send e-mail to majordomo@iss.net and, in the text of your message (not the subject line), write:

unsubscribe ntsecurity

This is an moderated mailing list discussing Windows NT security as well as the Windows 95 and Windows For Work Group security issues. The issues discussed will be everything at the host and application level security as well as at the network level.

Phrack

To join, send e-mail to phrack@well.com and, in the text of your message (not the subject line), write:

subscribe phrack

Phrack is a Hacker Magazine which deals with phreaking and hacking.

PRIVACY Forum

To join, send e-mail to privacy-request@vortex.com and, in the text of your message (not the subject line), write:

information privacy

The PRIVACY Forum is run by Lauren Weinstein. He manages it as a rather selectively moderated digest, somewhat akin to RISKS; it spans the full range of both technological and non-technological privacy-related issues (with an emphasis on the former).

Risks

To join, send e-mail to risks-request@csl.sri.com and, in the text of your message (not the subject line), write:

subscribe

Risks is a digest that describes many of the technological risks that happen in today's environment.

Secure HTTP

To join, send e-mail to shttp-talk-request@OpenMarket.com and, in the text of your message (not the subject line), write:

subscribe

Secure NCSA httpd is a World-Wide Web (WWW) server supporting transaction privacy and authentication for Secure WWW clients over the Internet using the Secure HyperText Transfer Protocol (S-HTTP). Secure NCSA httpd was developed by Enterprise Integration Technologies in cooperation with RSA Data Security and the National Center for Supercomputing Applications at the University of Illinois, Urbana-Champaign.

The purpose of this mailing list (shttp-talk) is to allow people who are interested in potentially using SHTTP to ask questions, air issues, express concerns and discuss the specification and reference implementation. Information about Secure HTTP can be found on the CommerceNet WWW server.

Sneakers

To join, send e-mail to majordomo@CS.YALE.EDU and, in the text of your message (not the subject line), write:

subscribe Sneakers

The Sneakers mailing list is for discussion of LEGAL evaluations and experiments in testing various Internet "firewalls" and other TCP/IP network security products.

Secure Socket Layer—Talk

To join, send e-mail to ssl-talk-request@netscape.com and, in the text of your message (not the subject line), write:

subscribe

Mailing list to discuss secure sockets layer - Netscape's (and, increasingly, others') approach to providing encryption and authentication for IP-based services (primarily http, but expanding to address telnet and ftp as well).

UNINFSEC—University Info Security Forum

To join, send e-mail to listserv@cuvmc.ais.columbia.edu and, in the text of your message (not the subject line), write:

subscribe uninfsec

This is a closed, unmoderated discussion list for people that have information security responsibilities in their jobs and who work for educational institutions or have a close relation with education. Discussions range from policy discussions, awareness programs, virus protection, change control, privileges, monitoring, risk assessments, auditing, business resumption, etc.

Virus

To join, send e-mail to LISTSERV@lehigh.edu and, in the text of your message (not the subject line), write:

subscribe virus-l your-name

It is an electronic mail discussion forum for sharing information and ideas about computer viruses, which is also distributed via the Usenet Netnews as comp.virus. Discussions should include (but not necessarily be limited to): current

events (virus sightings), virus prevention (practical and theoretical),
and virus related questions/answers. The list is moderated and digested. That means that any message coming in gets sent to me, the editor. I read through the messages and make sure that they adhere to the guidelines of the list (see below) and add them to the next digest. Weekly logs of digests are kept by the LISTSERV (see below for details on how to get them). For those interested in statistics, VIRUS-L is now up to about 2400 direct subscribers. Of those, approximately 10% are local redistribution accounts with an unknown number of readers. In addition, approximately 30,000–40,000 readers read comp.virus on the USENET.

Virus Alert

To join, send e-mail to LISTSERV@lehigh.edu and, in the text of your message (not the subject line), write:

subscribe valert-l your-name

What is VALERT-L?

It is an electronic mail discussion forum for sharing urgent virus warnings among other computer users. Postings to VALERT-L are strictly limited to warnings about viruses (e.g., "We here at University/Company X just got hit by virus Y—what should we do?"). Follow-ups to messages on VALERT-L should be done either by private e-mail or to VIRUS-L, a moderated, digested, virus discussion forum also available on this LISTSERV, LISTSERV@LEHIGH.EDU. Note that any message sent to VALERT-L will be cross-posted in the next VIRUS-L digest. To preserve the timely nature of such warnings and announcements, the list is moderated on demand (see posting instructions below for more information).

*What VALERT-L is *not*?*

A place to anything other than announce virus infections or warn people about particular computer viruses (symptoms, type of machine which is vulnerable, etc.).

WWW Security

To join, send e-mail to www-security-request@nsmx.rutgers.edu and, in the text of your message (not the subject line), write:

subscribe www-security your_e-mail_address

The list is maintained by the www-security team of Network Services, Rutgers University Telecommunications Division.

www-security is the official mailing list of the IETF Web Transaction Security Working Group. While there are many approaches to providing security services in the Web, most of the current work is concerned with securing the Hyper-Text Transport Protocol. Because of (1) the great need for quick implementation of Web security services, (2) HTTP-level solutions cover a wide range of WWW applications, and (3) the IETF is a proven forum for promoting standards to vendors and the international networking community, we suggest that the list focus and development of Internet standards and related documents for secure services within HTTP.

Vendors and Organizations

CERT (Computer Emergency Response Team) Advisory mailing list.

To join, send e-mail to cert@cert.org and, in the text of your message (not the subject line), write:

subscribe

CIAC (Computer Incident Advisory Capability) of the US Department of Energy.

CIAC manages the following mailing list for its electronic publications:

CIAC-Bulletin, CIAC Information Bulletins, and Advisory Notices containing important, time-critical computer security information.

To join, send e-mail to majordomo@tholia.llnl.gov and, in the BODY of your message (not the subject line), write any of the following examples:

subscribe ciac-bulletin

HP, Hewlett Packard

To join, send e-mail to support@support.mayfield.hp.com and, in the text of your message (not the subject line), write:

subscribe security_info

The latest digest of new HP Security Bulletins will be distributed directly to your mailbox on a routine basis.

Sun Security Alert

To join, send e-mail to security-alert@sun.com and, in the subject of your message write:

subscribe cws your-e-mail-addr

The message body should contain affiliation and contact information.

Reporting Computer Crimes

If you are the victim of any kind of computer crime or other violation in Cyberspace, report it. Make sure your documentation is as complete as possible, and be specific to make sure you speak to the right people who also speak tech jargon. If there's any question about who to report it to, call your local police department, your state police or the

FBI and ask for help. While most computer crimes do invite investigation, jurisdiction is sometimes confusing, so you may have to stick with it until you find the right group.

Computer Emergency Response Team—Coordination Center
Software Engineering Institute
Carnegie Mellon University
Pittsburgh, Penn. 15213
412-268-7090

A federally funded organization to reactively and proactively promote security. As a centralized repository for technical problems, system wide breaches or attacks, especially on the Internet, they can generate a national response in minutes.

Federal Bureau of Investigation - FBI
National Computer Crime Squad
J. Edgar Hoover Building,
10th and Pennsylvania Ave.
Washington, D.C., 20535.

For interstate crimes and those concerning 'federal interest computers', international incidents and computer crimes in general. If in doubt, call them, and they'll point you in the right direction. Contact your local bureau office; their number is in the phone book.

U.S. Secret Service
Electronic Crimes Branch
1800 G St. Room 900 Washington, D.C. 20223
202-435-5850

For computer and communications crimes involving federal interest computers, Presidential threats, counterfeiting and general computer intrusions. Does not investigate classified security breaches. Coordinates with the FBI for jurisdiction and often the two agencies share cases.

ISN (Information Security Network)

Subscribe to: majordomo@repsec.com

With message: subscribe isn

POLITECH (the moderated mailing list of politics and technology)

Subscribe to: majordomo@vorlon.mit.edu

With message: subscribe politech

CNET: The Digital Domain (great news site for wired culture news)
www.cnet.com

WIRED.COM: Electronic Homepage of Wired Magazine

www.wired.com

Extranet for Security Professionals (US Government sponsored)

www.xsp.org

Associated Press (AP) News Wire Feeds

wire.ap.org

G-TWO-I (Get The Word Out – Intelligence)

In 1996, a loose organization of national security academics and practitioners formed a mailing list called "g2i" which is a play on the military intelligence officer's designation and stands for "Get The Word Out – Intelligence." Since inception, g2i has grown to over three hundred members that exchange information related to their work supporting the national and community security of the United States. g2i members include Congressional staff, military forces stationed overseas and domestically, very senior members of the National Security community, state and local law enforcement, and subject-matter experts hand-picked from colleges around the country that provide detailed analytical support to the list. Membership on the list is restricted to military, government, select law enforcement and academic subject-matter experts.

g2i has been recognized by the national security community and the media as an innovative program to get critical unclassified information to those that really need it without having to go through the typical Washington layers of bureaucracy and secrecy associated with intelligence information. Often times members receive complements and letters from "subscribers" in the field. One letter came from an Army Special Forces officer stationed in the Middle East who uses g2i as his primary source of intelligence information about the areas he is stationed in as it is unclassified, easy to work with, and he can pose specific questions for further analysis. A member of the Joint Chiefs of Staff uses a civilian analyst to compare g2i analysis with that of large military organizations, and often uses g2i material to brief his general officer instead of government-generated information. Success stories like this have propelled g2i into the mainstream as a viable, accurate, timely, and reliable information source for our national security and local law enforcement professionals.

Information regarding this (and other) g2I services can be requested through info@taoiw.org.

The Z-GRAM

The ZGram is an electronic newsletter containing news, resources, products, symposia announcements, employment, and business opportunities regarding the United States Defense and Law Enforcement Communities. The ZGram is delivered Monday through Friday to readers via email.

The ZGram is distributed free of charge to members of the National Military Intelligence Association (www.nmia.org) and members of the Operations Security Professionals Society (www.opsec.org).

If you are not a member of either NMIA or OPS, you may request an annual subscription to the ZGram by forwarding US $98.00 (check made payable to Real Trends, Inc.) to ZGram, 9200 Centerway Road, Gaithersburg, Maryland 20879. Be sure to include your email address. Ensuring that you submit the correct email address to send the Zgram to—print clearly.

You may request a sample issue by sending email to zhi@zgram.net. Please write "ZGRAM SAMPLE" in the subject line.

As an extra service to subscribers, The ZGram Attic was created and contains a wealth of links and reference materials found while researching the ZGram. The Attic may be found at www.zgram.net.

EmergencyNet News Service

The EmergencyNet NEWS Service is the reporting arm and a subsidiary of the Chicago-based Emergency Response and Research Institute (ERRI). ERRI is an emergency service/military "think-tank" that studies evolving events, problems, and policy germane to the emergency response, intelligence, military, and national security communities. Both ERRI and the EmergencyNet News Service are nonpartisan, non-political, and completely independent and does not receive any funding or other influence from any political, ethnic, religious, or ideological group or organization.

ERRI has been monitoring threats to the national security and business communities and other interests since 1990. It published its findings and related materials through the EmergencyNet News service. EmergencyNet News specializes in providing timely, comprehensive, and objective reporting on international and domestic terrorism, crime, disasters, and related activities, as well as other emergency response topics.

EmergencyNet News generates a number of publications, special reports, flash messages and links to ERRI consulting services to accomplish our goal of providing the most factual, in-depth and timely information to our subscribers and clients, in order to help them maximize their knowledge and not fall victim to the dangers associated with crime, terrorism and other violence or disasters. EmergencyNet News is also distributed to a number of national and international news agencies, TV stations, Radio stations, emergency magazines, and other periodicals.

ERRI publications include the ENN Daily Intelligence Report, the ENN Daily Emergency Services Report, the ERRI World Situation Report and the ENN Weekly Watch Report. ENN reporting is available through the on-line EMERGENCY electronic bulletin board, or by e-mail. In addition to studies and reports, ERRI conducts consulting and education programs in a large number of emergency-related areas, including specialized consulting for emergency response agencies, major corporations, small and medium-sized businesses, financial institutions, law firms and government entities.

Emergency Response and Research Institute
6348 N Milwaukee Ave, Suite 312, Chicago,
Illinois 60646 USA
773-631-ERRI Voice/Voice Mail
773-631-4703 Fax
Main WWW page: http://www.emergency.com

Information Warfare Glossary

(Department of Defense interpretations)

civil affairs. The activities of a commander that establish, maintain, influence, or exploit relations between military forces and civil authorities, both governmental and nongovernmental, and the civilian populace in a friendly, neutral, or hostile area of operations in order to facilitate military operations and consolidate operational objectives. Civil affairs may include performance by military forces of activities and functions normally the responsibility of local government. These activities may occur prior to, during, or subsequent to other military actions. They may also occur, if directed, in the absence of other military operations. Also called CA.

command and control. The exercise of authority and direction by a properly designated commander over assigned and attached forces in the accomplishment of the mission. Command and control functions are performed through an arrangement of personnel, equipment, communications, facilities, and procedures employed by a commander in planning, directing, coordinating, and controlling forces and operations in the accomplishment of the mission. Also called C2.

command and control warfare. The integrated use of operations security, military deception, psychological operations, electronic warfare, and physical destruction, mutually supported by intelligence, to deny information to, influence, degrade, or destroy adversary command and control capabilities, while protecting friendly command and control capabilities against such actions. Command and control warfare is an application of information operations in military operations. Also called C2W. C2W is both offensive and defensive:

C2-attack. Prevent effective C2 of adversary forces by denying information to, influencing, degrading, or destroying the adversary C2 system.

C2-protect. Maintain effective command and control of own forces by turning to friendly advantage or negating adversary efforts to deny information to, influence, degrade, or destroy the friendly C2 system.

communications security. The protection resulting from all measures designed to deny unauthorized persons information of value which might be derived from the possession and study of telecommunications, or to mislead unauthorized persons in their interpretation of the results of such possession and study. Also called COMSEC. Communications security includes cryptosecurity, transmission security, emission security, and physical security of communications security materials and information.

> ***cryptosecurity***—The component of communications security that results from the provision of technically sound cryptosystems and their proper use.
>
> ***transmission security***—The component of communications security that results from all measures designed to protect transmissions from interception and exploitation by means other than cryptanalysis
>
> ***emission security***—The component of communications security that results from all measures taken to deny unauthorized persons information of value that might be derived from intercept and analysis of compromising emanations from crypto-equipment and telecommunications systems.
>
> ***physical security***—The component of communications security that results from all physical measures necessary to safeguard classified equipment, material, and documents from access thereto or observation thereof by unauthorized persons.

computer network attack. Operations to disrupt, deny, degrade, or destroy information resident in computers and computer networks, or the computers and networks themselves. Also called CNA.

computer security. The protection resulting from all measures to deny unauthorized access and exploitation of friendly computer systems. Also called COMPUSEC.

counterintelligence. Information gathered and activities conducted to protect against espionage, other intelligence activities, sabotage, or assassinations conducted by or on behalf of foreign governments or elements thereof, foreign organizations, or foreign persons, or international terrorist activities. Also called CI or internal security.

deception. Those measures designed to mislead the enemy by manipulation, distortion, or falsification of evidence to induce him to react in a manner prejudicial to his interests.

defensive information operations. The integration and coordination of policies and procedures, operations, personnel, and technology to protect and defend information and information systems. Defensive information operations are conducted through information assurance, physical security, operations security, counter-deception, counter-psychological operations, counterintelligence, electronic warfare, and special information operations. Defensive information operations ensure timely, accurate, and relevant information access while denying adversaries the opportunity to exploit friendly information and information systems for their own purposes.

directed-energy warfare. Military action involving the use of directed-energy weapons, devices, and countermeasures to either cause direct damage or destruction of enemy equipment, facilities, and personnel, or to determine, exploit, reduce, or prevent hostile use of the electromagnetic spectrum through damage, destruction, and disruption. It also includes actions taken to protect friendly equipment, facilities, and personnel and retain friendly use of the electromagnetic spectrum. Also called DEW.

electronic warfare. Any military action involving the use of electromagnetic and directed energy to control the electromagnetic spectrum or to attack the enemy. Also called EW. The three major subdivisions within electronic warfare are: electronic attack, electronic protection, and electronic warfare support.

electronic attack. That division of electronic warfare involving the use of electromagnetic, directed energy, or antiradiation weapons to attack personnel, facilities, or equipment with the intent of degrading, neutralizing, or destroying enemy combat capability. Also called EA. EA includes: 1) actions taken to prevent or reduce an enemy's effective use of the electromagnetic spectrum, such as jamming and electromagnetic deception, and 2) employment of weapons that use either electromagnetic or directed energy as their primary destructive mechanism (lasers, radio frequency weapons, particle beams, or antiradiation weapons).

electronic protection. That division of electronic warfare involving actions taken to protect personnel, facilities, and equipment from any effects of friendly or enemy employment of electronic warfare that degrade, neutralize, or destroy friendly combat capability. Also called EP.

electronic warfare support. That division of electronic warfare involving actions tasked by, or under direct control of, an operational commander to search for, intercept, identify, and locate sources of intentional and unintentional radiated electromagnetic energy for the purpose of immediate threat recognition. Thus, electronic warfare support provides information required for immediate decisions involving electronic warfare operations and other tactical actions such as threat avoidance, targeting, and homing. Also called ES. Electronic warfare support data can be used to produce signals intelligence, both communications intelligence, and electronic intelligence.

global information infrastructure. The worldwide interconnection of communications networks, computers, databases, and consumer electronics that make vast amounts of information available to users. The global information infrastructure encompasses a wide range of equipment, including cameras, scanners, keyboards, facsimile machines, computers, switches, compact disks, video and audio tape, cable, wire, satellites, fiber-optic transmission lines, networks of all types, televisions, monitors, printers, and much more. The friendly and adversary personnel who make decisions and handle the transmitted information constitute a critical component of the global information infrastructure. Also called GII.

incident. In information operations, an assessed event of attempted entry, unauthorized entry, or an information attack on an automated information system. It includes unauthorized probing and browsing; disruption or denial of service; altered or destroyed input, processing, storage, or output of information; or changes to information system hardware, firmware, or software characteristics with or without the users' knowledge, instruction, or intent.

indications and warning. Those intelligence activities intended to detect and report time-sensitive intelligence information on foreign developments that could involve a threat to the United States or allied/coalition military, political, or economic interests or to US citizens abroad. It includes forewarning of enemy actions or intentions; the imminence of hostilities; insurgency; nuclear/non-nuclear attack on the United States, its overseas forces, or allied/coalition nations; hostile reactions to US reconnaissance activities; terrorists' attacks; and other similar events. Also called I&W.

information. 1. Facts, data, or instructions in any medium or form. 2. The meaning that a human assigns to data by means of the known conventions used in their representation.

information assurance. Information operations that protect and defend information and information systems by ensuring their availability, integrity, authentication, confidentiality, and non-repudiation. This includes providing for restoration of information systems by incorporating protection, detection, and reaction capabilities. Also called IA.

information environment. The aggregate of individuals, organizations, or systems that collect, process, or disseminate information; also included is the information itself.

information operations. Actions taken to affect adversary information and information systems while defending one's own information and information systems. Also called IO.

information security. Information security is the protection and defense of information and information systems against unauthorized access or modification of information, whether in storage, processing, or transit, and against denial of service to authorized users. Information security includes those measures necessary to detect, document, and counter such threats. Information security is composed of computer security and communications security. Also called INFOSEC.

information superiority. The capability to collect, process, and disseminate an uninterrupted flow of information while exploiting or denying an adversary's ability to do the same.

information system. The entire infrastructure, organization, personnel, and components that collect, process, store, transmit, display, disseminate, and act on information.

information warfare. Information operations conducted during time of crisis or conflict to achieve or promote specific objectives over a specific adversary or adversaries. Also called IW.

intelligence preparation of the battlespace. An analytical methodology employed to reduce uncertainties concerning the enemy, environment, and terrain for all types of operations. Intelligence preparation of the battlespace builds an extensive data base for each potential area in which a unit may be required to operate. The data base is then analyzed in detail to determine the impact of the enemy, environment, and terrain on operations and presents it in graphic form. Intelligence preparation of the battlespace is a continuing process. Also called IPB. In the commercial world, it is called "proper planning."

leveraging. In information operations, the effective use of information, information systems, and technology to increase the means and synergy in accomplishing information operations strategy.

military deception. Actions executed to deliberately mislead adversary military decision-makers as to friendly military capabilities, intentions, and operations, thereby causing the adversary to take specific actions (or inactions) that will contribute to the accomplishment of the friendly mission. The five categories of military deception are:

strategic military deception—Military deception planned and executed by and in support of senior military commanders to result in adversary military policies and actions that support the originator's strategic military objectives, policies, and operations.

operational military deception—Military deception planned and executed by and in support of operational-level commanders to result in adversary actions that are favorable to the originator's objectives and operations. Operational military deception is planned and conducted in a theater of war to support campaigns and major operations.

tactical military deception—Military deception planned and executed by and in support of tactical commanders to result in adversary actions that are favorable to the originator's objectives and operations. Tactical military deception is planned and conducted to support battles and engagements.

Service military deception—Military deception planned and executed by the Services that pertain to Service support to joint operations. Service military deception is designed to protect and enhance the combat capabilities of Service forces and systems.

military deception in support of operations security (OPSEC)—Military deception planned and executed by and in support of all levels of command to support the prevention of the inadvertent compromise of sensitive or classified activities, capabilities, or intentions. Deceptive OPSEC measures are designed to distract foreign intelligence away from, or provide cover for, military operations and activities.

military operations other than war. Operations that encompass the use of military capabilities across the range of military operations short of war. These military actions can be applied to complement any combination of the other instruments of national power and occur before, during, and after war. Also called MOOTW.

national information infrastructure. The nation-wide interconnection of communications networks, computers, databases, and consumer electronics that make vast amounts of information available to users. The national information infrastructure encompasses a wide range of equipment, including cameras, scanners, keyboards, facsimile machines, computers, switches, compact disks, video and audio tape, cable, wire, satellites, fiber-optic transmission lines, networks of all types, televisions, monitors, printers, and much more. The friendly and adversary personnel who make decisions and handle the transmitted information constitute a critical component of the national information infrastructure. Also called NII.

offensive information operations. The integrated use of assigned and supporting capabilities and activities, mutually supported by intelligence, to affect adversary decisionmakers to achieve or promote specific objectives. These capabilities and activities include, but are not limited to, operations security, military deception, psychological operations, electronic warfare, physical attack and/or destruction, and special information operations, and could include computer network attack

operational level of war. The level of war at which campaigns and major operations are planned, conducted, and sustained to accomplish strategic objectives within theaters or areas of operations. Activities at this level link tactics and strategy by establishing operational objectives needed to accomplish the strategic objectives, sequencing events to achieve the operational objectives, initiating actions, and applying resources to bring about and sustain these events. These activities imply a broader dimension of time or space than do tactics; they ensure the logistic and administrative support of tactical forces, and provide the means by which tactical successes are exploited to achieve strategic objectives.

operations security. A process of identifying critical information and subsequently analyzing friendly actions attendant to military operations and other activities to:

Identify those actions that can be observed by adversary intelligence systems.

Determine indicators hostile intelligence systems might obtain that could be interpreted or pieced together to derive critical information in time to be useful to adversaries.

Select and execute measures that eliminate or reduce to an acceptable level the vulnerabilities of friendly actions to adversary exploitation. Also called OPSEC. In the corporate or commercial world, known as "common sense."

perception management. Actions to convey and/or deny selected information and indicators to foreign audiences to influence their emotions, motives, and objective reasoning; and to intelligence systems and leaders at all levels to influence official estimates, ultimately resulting in foreign behaviors and official actions favorable to the originator's objectives. In various ways, perception management combines truth projection, operations security, cover and deception, and psychological operations.

physical security. That part of security concerned with physical measures designed to safeguard personnel; to prevent unauthorized access to equipment, installations, material, and documents; and to safeguard them against espionage, sabotage, damage, and theft.

probe. In information operations, any attempt to gather information about an automated information system or its on-line users.

psychological operations. Planned operations to convey selected information and indicators to foreign audiences to influence their emotions, motives, objective reasoning, and ultimately the behavior of foreign governments, organizations, groups, and individuals. The purpose of psychological operations is to induce or reinforce foreign attitudes and behavior favorable to the originator's objectives. Also called PSYOP. In the commercial world, it is referred to as "advertising."

public affairs. Those public information, command information, and community relations activities directed toward both the external and internal publics with interest in the Department of Defense. Also called PA.

special information operations. Information operations that by their sensitive nature, due to their potential effect or impact, security requirements, or risk to the national security of the United States, require a special review and approval process. Also called SIO.

strategic level of war. The level of war at which a nation, often as a member of a group of nations, determines national or multinational (alliance or coalition) security objectives and guidance, and develops and uses national resources to accomplish these objectives. Activities at this level establish national and multinational military objectives; sequence initiatives; define limits and assess risks for the use of military and other instruments of national power; develop global plans or theater war plans to achieve these objectives; and provide military forces and other capabilities in accordance with strategic plans.

tactical level of war. The level of war at which battles and engagements are planned and executed to accomplish military objectives assigned to tactical units or task forces. Activities at this level focus on the ordered arrangement and maneuver of combat elements in relation to each other and to the enemy to achieve combat objectives.

vulnerability. 1. The susceptibility of a nation or military force to any action by any means through which its war potential or combat effectiveness may be reduced or its will to fight diminished. 2. The characteristics of a system which cause it to suffer a definite degradation (incapability to perform the designated mission) as a result of having been subjected to a certain level of effects in an unnatural (manmade) hostile environment, and 3. In information operations, a weakness in information system security design, procedures, implementation, or internal controls that could be exploited to gain unauthorized access to information or an information system.

vulnerability analysis. In information operations, a systematic examination of an information system or product to determine the adequacy of security measures, identify security deficiencies, provide data from which to predict the effectiveness of proposed security measures, and confirm the adequacy of such measures after implementation.

ABOUT THE AUTHORS

Richard Forno is currently the Director of Security for a major internet services firm in Herndon, VA. His recent activities include helping establish the very active Information Resources Security Office at the US House of Representatives and serving as a special consultant to the Office of the Secretary of Defense. His major areas of interest include national information warfare issues, computer crime investigations, and security education.

Richard is a graduate of Valley Forge Military College, The American University School of International Service, and is the youngest graduate on record from the United States Naval War College. He is a frequent speaker and author on the subjects of information warfare and security management, with articles and commentary featured in several publications including *Federal Computer Week, Technology Week, Forbes, and the Journal of Operations Security.*

Contact him at <rforno@taoiw.org>

Ronald Baklarz is currently the Director of Network Monitoring and Incident Handling for a major insurance firm in Newark, NJ. His recent activities include building information security programs within the Naval Nuclear Program and at the US House of Representatives. His major areas of interest include network intrusion detection, computer-assisted pattern recognition, and cyber-forensics.

Ron holds a Master of Science degree in Information Science and a Certificate of Advanced Study in Telecommunications both from the University of Pittsburgh. He is also a Certified Information Systems Security Professional (CISSP) and frequent speaker and author on the subjects of information warfare and security management.

Contact him at <rbaklarz@taoiw.org>

For additional information on this book,
its authors, or a calendar of appearances, lectures, or book signings,
please visit www.taoiw.org